Stand-Alone Photovoltaic Applications

LESSONS LEARNED

Technical editors:

GEERLING LOOIS AND BERNARD VAN HEMERT

Edited by

Published by
James & James (Science Publishers) Ltd,
35–37 William Road, London NW1 3ER, UK

This publication was sponsored by the DG XII of the European Commission
under its Joule programme, contract JOR3–CT97–2004

ISBN 1 873936 91 5

Printed in Hong Kong by Magnum International Printing Co. Ltd

Cover photographs courtesy Trama TecnoAmbiental and H. de Gooyer

Contents

Foreword

One of the important goals of the IEA Photovoltaic Power Systems (PVPS) Programme is to share 'Lessons learned' about various applications of PV systems, not only among experts of the participating countries, but also with interested parties around the world.

Task III of the PVPS programme, active since 1993, focuses on the exchange of information on stand-alone PV applications. Stand-alone PV systems will continue to represent a significant PV market segment, not only in developing countries, but also in the important home markets of industrial countries. Experts from 15 countries are sharing their experiences in this area.

This publication is a typical example of fruitful international cooperation, and contains a variety of lessons learned. They demonstrate the influence of local circumstances and the different approaches towards PV taken in different countries. The lessons are presented in a generalised form, and are already being incorporated in new stand-alone applications and projects all over the world.

I trust that this publication will contribute to a wider and better use of PV systems, servicing vital needs of end-users.

Erik H. Lysen
Chairman Executive Committee
IEA-PVPS Programme
Utrecht, The Netherlands
February 1999

Acknowledgements

This publication was made possible thanks to the efforts and contributions of many individuals and organisations.

Fifteen authors from as many countries contributed their sections, and they were given full support from their respective organisations. We would like to thank them for the patience and conscientiousness with which they wrote and re-wrote their sections, until all contributions would finally fit an overall framework.

This publication would not have materialised without a financial contribution from the European Commission (DG XII, JOR3-CT97-2004) towards the editing, for which support we want to express our gratitude.

Apart from these key players, we should not forget those co-authors, reviewers, secretaries and technicians who all contributed their specific expertise to make this book what it is.

List of abbreviations and acronyms

AC   Alternating Current

BOS   Balance Of System, all components of an SAPV system that are not part of the array subsystem, i.e. the power conditioning, energy storage, power distribution and control

DC   Direct Current

DG   Diesel Generator

DOD   Depth of Discharge: Ampere-hours removed from a charged cell or battery in one discharge period. Is often expressed in as a percentage of rated capacity. See also section 7.3

EC   European Commission

EPIA European PV Industry Association

ESCO Energy Service Company, see section 4.3

IEA   International Energy Agency

MPPT Maximum Power Point Tracking, electronic control device aiming at operating the array at its maximum power point. Applied only in relatively large systems. See section 7.1.

NGO   Non-Governmental Organisation

O&M   Operation and Maintenance

PL   Light, energy-efficient light commonly used in PV systems

PMC   Product-Market Combination

PV   PhotoVoltaic; the direct conversion of light into electricity

PVPS Photovoltaics Power Systems Programme from the IEA

RO   Reverse Osmosis: filter technology, see Australian showcase

RD&D Research, Development and Demonstration

SAPV Stand-Alone PhotoVoltaic, see Box 1.2

SOC   State of Charge of a battery. See section 7.3

Wp   Watt-peak or peak-watt, the maximum yield of an array under Standard Test Conditions. Indicates the size of a PV generator. Also encountered as kWp, MWp, GWp

Notes

1   Terminology in the PV technology is not uniform and contains a high degree of jargon. The same word can have different connotations in different circles, while translations to other languages can further increase the differences. We have tried to adopt the terminology as propagated by the EC and laid down in its handbook *Photovoltaic System Technology* [1].

2   Electro-technical entities such as Volts, Ampères, Watts etc are not included.

Part I

Lessons learned in SAPV applications

1 Introduction

1.1 THE INTERNATIONAL ENERGY AGENCY

The International Energy Agency was established in 1974 as an autonomous agency within the Organisation for Economic Cooperation and Development (OECD). The European Commission also participates in the work of the agency. The IEA carries out a comprehensive programme of energy cooperation among its 23 member countries. The IEA Photovoltaic Power Systems Programme (PVPS) is one of the collaborative R&D agreements established within the IEA, and its participants have conducted a variety of joint projects in the applications of photovoltaics.

The overall programme is headed by an Executive Committee composed of one representative from each participating country, while the management of individual research projects (Tasks) is the responsibility of Operating Agents. Currently, seven tasks have been established.

The objective of Task III is to promote and facilitate the exchange of information and experiences in the field of PV Systems in Stand-alone and Island Applications. Prominent activities performed in Task III are:

- support to international organisations (e.g. the World Bank) regarding SAPV applications in developing countries, especially solar home systems;
- production of a series of Practical Recommendations for all system components;
- exchange of information on problems and solutions in national projects, resulting in this publication.

1.2 AIM: TO CONSOLIDATE THE LESSONS LEARNED

The participating countries in Task III have shared and brought together a wide range of experiences regarding all aspects of the implementation of stand-alone PV systems. Not only has a lot of knowledge been gained regarding the technical aspects; considerable progress has also been made in the economic, social, organisational and marketing aspects.

This book aims to illustrate and consolidate this progress by presenting a selection of lessons learned. These should in turn contribute to improved quality of current and future systems and projects, like the national showcase projects of Part II.

The book focuses on the practical experiences gained, and does not aim to provide a complete manual on SAPV. When Task III started its activities in 1993, a collection of 50 'State of the art' projects was published in the book *Examples of Stand-Alone Photovoltaic Systems* [2]. This publication marked the base line for the work of the task. Now, in 1998, the showcases from each country demonstrate the lessons learned in five years of cooperation.

To this end, experts from all 15 member countries (see Box 1.1) contributed to the main text and the showcases, with funding by their respective national institutions. The editing and layout was done by ECOFYS in The Netherlands with financial support from DG XII from the European Commission.

1.3 STRUCTURE OF THE BOOK

The book consists of two parts. The first part contains eight chapters dealing with a specific aspect of stand-alone PV. The second part introduces 14 national showcase projects in a systematic presentation. The design is such that each chapter and showcase can be read independently from the rest of the book.

Chapter 2, contributed by The Netherlands, analyses *the market* for stand-alone PV systems. It gives an overview of the 'traditional' application of stand-alone PV, which is the electrification of remote buildings and which has been addressed in depth in other publications. The focus is on the market niches of service applications that are also interesting for more densely populated areas, e.g. in industrialised countries.

The United Kingdom illustrates *the economic aspects* in *Chapter 3*. Cost comparisons are made, but more important is the illustration of the non-financial considerations that make PV the preferred choice as a power source for many applications.

Switzerland explores in *Chapter 4 (financing aspects)* different financing mechanisms and financial policies used to overcome the initial cost barrier. Most of these approaches have been applied in developing countries rather than in the western world.

Using various examples from all over the world, France analyses in *Chapter 5* the *institutional aspects*: what are the different roles for all possible actors, not only in the planning and implementation phases but also in the after-

Box 1.1: Experts participating in Task III and authors of showcase projects

Country	Expert	Address
Australia	Keith Presnel	Power and Water Authority
		fax +61-(0) 88 92 47 430, e-mail keith.presnell@nt.gov.au
Canada	Sylvain Martel	CANMET-EDRL
		fax +1-(0) 450 652 5177, e-mail: sylvain.martel@nrcan.gc.ca
Finland	Lauri Manninnen	Neste OY NAPS
		fax +358-(0) 20 45 07 113, e-mail: lauri.manninen@neste.com
France	Philippe Malbranche	GENEC
		fax +33-(0) 44 22 57 365, e-mail: philippe.malbranche@cea.fr
Germany	Ingo Stadler	University of Kassel
		fax+49-(0) 561 88 04 64 34
		e-mail: stadler@re.e-technik.uni-kassel.de
Italy	Carlo Zuccaro	ENEL
		fax +39-(0) 27 22 45 465, e-mail: zuccaro@dsr.enel.it
Japan	Yoshiyuki Ishihara	Doshisha University
		fax +81-(0) 77 46 56 813, e-mail: yishihar@mail.doshisha.ac.jp
Korea	Man Geun Lee	KIER
		fax +82-(0) 42 86 03 739, e-mail: l-mg@sun330.kier.re.kr
Netherlands	Geerling Loois	ECOFYS
		fax +31-(0)30 2808301, e-mail: G.Loois@ecofys.nl
Norway	Knut Hofstad	NVE
		fax +47-(0) 22 95 90 71, e-mail: kho@nve.no
Portugal	Antonio Joyce	INETI
		fax +351-(0) 17 16 51 141, e-mail: antonio.Joyce@ite.ineti.pt
Spain	Xavier Vallvé	TTA
		fax +34-(0) 93 45 66 948, e-mail: tta@mx3.redestb.es
Sweden	Bengt Perers	Vattenfall Utveckling,
		fax: +46-(0) 15 52 29 30 60,
		e-mail: bengt.perers@utveckling.vattenfall.se
Switzerland	Bernard Bezençon	Atlantis Solar Systems
		fax: +41-(0) 31 30 03 220, e-mail: Atlantis@access.ch
United Kingdom	Alison Wilshaw	IT Power
		fax: +44-(0) 11 89 73 08 20, e-mail: arw@itpower.co.uk

For some countries, the main author or co-author of the showcase was a different person:

Country	Author	Address
France	Patrick Jourde	GENEC
		fax +33 442 25 73 65, e-mail: jourde@macadam.cea.fr
Portugal	Carlos Rodriguez	INETI
		fax (351) 17 16 51 141, e-mail: carlos.rodriguez@ite.ineti.pt

Box 1.2: Terminology and concepts

This box presents some basic terminology frequently used in this book:

- *Stand-alone Photovoltaics (SAPV)*, also referred to as autonomous systems: applications powered by photovoltaics that operate independently of any electricity grid. The PV generator runs the application and recharges storage batteries, which in turn power the application when there is not enough insolation. If there is a surplus of solar power and the batteries are fully charged, this power is lost.
- *Grid-connected Photovoltaics*: PV generators directly connected to the electricity grid. The system can supply any surplus of energy to the grid, or extract energy if the PV generator does not meet the demand.

The broad field of stand-alone PV is divided into three applications (these categories are further elaborated in Chapter 2):

- *Service applications*, which include energy services in integrated applications such as telecommunications, water pumping, remote sensing, medical equipment, parking meters, bus stops, etc.
- *Remote buildings*, such as farms, houses, hotels, education centres
- *Island systems*; these larger PV-diesel systems are mostly located on islands, but also in remote villages, farms or commercial premises.

Under remote buildings, a sub-category of very basic PV systems can be distinguished, mostly referred to as:

- *Solar Home Systems*, basic PV systems, with generally around 50 Wp (generally not more than 200 Wp), that provide just enough energy for a few PL bulbs and some limited use of radio or TV. They are mostly applied in electrification of rural areas in developing countries and leisure cottages.

sales area. Legal aspects and the need for international standards are also discussed.

In *Chapter 6*, on *social aspects*, Spain uses the Garrotxa rural electrification project to demonstrate the importance of user involvement in all stages of the project cycle.

Chapter 7, written by Canada and Australia, gives an insight into the *technical aspects* of stand-alone PV.

In *Chapter 8*, The Netherlands looks at the future: starting from the *lessons* we have *learned*, and the main challenges of the next decade are explored in order to open gigawatt markets for SAPV.

The showcases of the second part are grouped according to the three categories: (service applications, remote buildings and island systems), and thereafter presented alphabetically by country.

We wish you much pleasure in reading this publication, and welcome any feedback.

ECOFYS, February 1999
Bernard van Hemert, Editor
Geerling Loois, Head SAPV

2 The market for stand-alone PV systems

Geerling Loois

ECOFYS, PO Box 8408, NL-3503 RK Utrecht, The Netherlands

This chapter gives an overview of SAPV markets and their potentials, barriers and opportunities. It concludes with lessons learned and research, development and demonstration (RD&D) needed to improve the market-oriented development of stand-alone PV systems.

2.1 OVERVIEW

At present most PV systems are stand-alone PV systems (SAPV), both in terms of number and in terms of installed PV power. This situation is expected to last for the coming decade. Currently, domestic markets for stand-alone PV systems in industrialized countries in Europe, the USA and Asia [3] represent at least half of the world-market.

This offers important opportunities for further development of the PV sector in these regions, since domestic markets represent a major source of income for the PV and related conventional industries. At the same time these markets offer a steep learning curve since customers, suppliers, facilitators and RD&D experts operate closely together.

The general expectation is that the current exponential growth of the PV market will continue. However, relative market shares will slowly change over a few decades.

- Stand-alone PV systems for rural electrification in developing countries are expected to experience an accelerated growth towards a large, commercially feasible market in the coming years.
- At the same time, PV in conventional products and services will continue to grow at a steady pace.
- An even stronger growth of PV in grid-connected applications is expected due to a decrease of PV systems costs.

Development of the market for professional PV systems will, in the short term, still lean heavily on private initiatives in the SAPV sector, partly supported by governments. This means that private market-oriented assignments will gain importance relative to traditional public funding. Stand-alone PV systems, products and services are still at an early phase of development, and end-users pose unsolved questions on, for instance, new materials and equipment, reliability, quality, standards, financing and market development.

2.2 MARKET POTENTIAL OF STAND-ALONE PV SYSTEMS

Stand-alone PV systems represent about 90% of today's total installed PV power. Up to 1997 about 640 MWp of PV had been installed worldwide [4]. In the SAPV market, three fields of application (communication, water pumping and domestic power supply) represent about 80% of the total installed SAPV power [5].

For the year 2010 a scenario of the European PV Industry Association (EPIA) forecasts a share for SAPV of 3 000 MWp (about 70% of the PV power by then installed). This view complies with the findings of a scenario study forecasting 700 MWp of SAPV in the European market in 2010 [6]. Similarly, UPVG expects for the USA a SAPV potential of 350 MWp [7]. This would imply a world-wide turnover of about US $ 30 billion in the next 15 years, generating a driving force for both the PV industry and market-oriented RD&D.

2.3 MARKET POTENTIAL PER APPLICATION

The three application groups as introduced in Box 1.2 (service applications, island systems and remote buildings) are described briefly in terms of applications, potentials, market parties and major needs.

Service applications

Service applications provide energy services, usually for appliances with a low energy demand. Typical proven applications are telecommunication, buoys and warning lights, water pumps and water management systems. Emerging markets are public lighting, transportation such as solar boats, cooling and ventilation and systems for monitoring, control and information exchange.

Extrapolation from Dutch and US studies shows that the commercially accessible potential in the next decade may be in the range of 1 GWp worldwide [8].

These energy services are even applicable in areas with a dense electricity grid. In most cases a PV system of 10–300 Wp is suitable. Exceptions only occur in regions with low irradiation, like the Polar Regions where larger PV arrays are used. Sometimes telecom relay stations, cathodic protection, water pumps and solar boats have a need for systems with a nominal power over 1 kWp.

Figure 2.1: Service applications: weir at Bunnik (The Netherlands). Photo: ECOFYS.

The low power demand and the consequent small PV systems make these stand-alone energy services competitive with grid extensions and connections, as will be elaborated in the next chapter.

The market potential can be enlarged when more emphasis is placed on energy efficiency. The conventional design of most energy services has not yet been optimized in this respect. Improvements in design enable reduction in energy demand by a factor of three to ten, especially in electrotechnical installations for energy services. This reduction lowers the systems costs and increases the competitiveness with the public grid.

The market for service applications is driven more by conventional industry than by the PV industry. In search of new products and independence from the electricity grid, innovative industries take the initiative to integrate PV systems into their products and services. This integration is usually accomplished in a joint effort by the PV and conventional industries.

In the development process of such PV systems, requirements are mostly subject to the specifications imposed by the conventional application industry. This has its effects not only on design, but also on technical product design. Developers tend to reduce the number of components and costs by combining functions. An illustrative example is daylight sensing through the PV array in PV streetlights. Another crucial point is the improvement in terms of energy efficiency, as this will significantly broaden the market.

Power supply for remote buildings

Power systems for remote buildings range from small SHS to large PV installations operating in a hybrid system with a diesel back-up generator. The installed power usually ranges from 50 to 2 000 Wp depending on the user needs. In exceptional cases the nominal PV power may be as high as 6 kWp.

The worldwide potential in this market exceeds 10 GWp [9]. Assuming an implementation period of 40 years to implement this, a potential of about 1 GWp will be installed before the year 2010.

In industrialized countries the main initiators in the market are the end-users and the utilities. In developing countries governmental authorities also take initiatives in the framework of social programmes. Utilities and authorities are discovering that SAPV is a cost-effective alternative to grid-connected rural electrification, both in industrialized and developing countries. For a full-scale market introduction appropriate schemes for (local) sales,

Figure 2.2: Remote buildings. Italy, 5–6 kWp stand-alone. Photo: ENEL.

quality control and service and financing are still under development [10,11].

In several countries initiatives for SAPV introduction are being taken. In Italy, The Netherlands and the USA utilities have developed schemes for 'remote pricing', in which customers are charged for the service instead of the amount of energy [12].

In Catalonia, Spain, innovative schemes of management and maintenance successfully rely on end-user involvement in user groups [13]. Similar schemes may be suitable for other areas.

Larger island systems

These systems are meant to supply electricity to islands, small villages (10 to 100 households) and small enterprises (like gas stations, recreational parks and hotels). Applications can overlap with service applications mentioned above. These systems typically combine 1 to 100 kWp of PV with an auxiliary diesel or wind generator of a few kW. The largest systems go up to 1 MWp PV with 1 MW diesel generators [14]. PV systems are also successfully added to existing generators, in order to improve the quality of the energy supply.

The world potential is estimated to be 0.5–1 GWp over the next two decades. An important market for these systems can be found in rural areas in developing countries and in the Pacific region with its numerous islands. For Europe a commercial potential of about a hundred systems per year (i.e. 1 MWp/year) is estimated [15].

Currently utilities and government organisations are the main players initiating projects for larger PV/hybrid systems. If these applications follow the trend of smaller service applications, commercial/private initiatives can be expected soon.

This application is least developed on a market and a systems level. This means that many questions (on sizing, economics, power and load management, management of the communal power supply, individual domestic power consumers, etc.) have to be resolved in order to move the application forward.

Figure 2.3: Island stems: village water supply in Algarve (Portugal). Photo: INETI.

2.4   BARRIERS

Apart from the financial limitations, four barriers to the large-scale introduction of stand-alone PV can be identified:

Lack of knowledge The main barrier is lack of knowledge. Apart from a few, one decade old, niche markets, the PV sector and conventional industry have exchanged little knowledge concerning each other's products, services and requirements. Even on a strategic level, assessments of high-potential combinations of products and markets hardly exist.

Box 2.1: SAPV market introduction strategy in The Netherlands

A Dutch strategic study showed that penetration of the Dutch market for stand-alone PV systems requires a coherent policy of information dissemination, creation of confidence, availability of attractive reliable products and policies of systematic integration of PV by large professional end-users.

Five product–market combinations (PMCs) have sufficient potential to reach an annual turnover in the Dutch market of 1 MWp by the year 2000.

Markets with a large potential in the short term are:

Light	Public lighting for end-users like local authorities, utilities and public transport;
Water	Measuring and monitoring equipment and pumps for the management of drinking and surface water for governmental organisations and agricultural enterprises;
Navigation	PV-electric propulsion of ships for professional and recreational use.

PMCs with a smaller commercial potential in the short term are:

Information	Communication and information points, ticket vending machines for public transport;
Cooling	Integration of PV power supply in cooling and ventilation systems for the private recreational market and the car industry.

In these five PMCs, a commercial potential between 2 and 6 MWp is expected in the coming few years, which is approximately two thirds of the estimated commercial potential for proven products with stand-alone PV supply in existing Dutch markets. As these PMCs are geographically evenly spread over The Netherlands, there is an opportunity to demonstrate these highly visible applications close to the end-users.

Based on an analysis of bottle-necks and solutions, an introduction programme is proposed for a period of four years, aiming at about 4 MWp (about 40 000 systems) in three to five PMCs.

The total turnover in complete products and systems for this programme is expected to be US $ 70 million, including costs of supporting measures and demonstration. The costs of the demonstration projects are expected to be supported mainly by the end-user. This is due to the fact that many stand-alone PV systems offer economic advantages, e.g. in comparison to grid extension and personnel costs.

The ALTENER programme of EC DG XVII (contract nr: AL/37/95/NI) and Novem financed the study[16].

Lack of confidence In niche markets where active knowledge transfer is taking place, both investors and clients often need to be convinced of the reliability of PV systems. This usually takes a period of several years, during which applications can only be demonstrated in small-scale projects. Also the absence of international standards hampers increase in confidence from newcomers.

Lack of products and services The range of available products is too narrow to respond fully to the end-users' needs. This not only concerns the products in terms of hardware and supply conditions. It also concerns accompanying services like financing, guarantees, etc.

Lack of integration in policies In the few market niches that have been developed, the previously mentioned barriers have been (partly) solved. This then results in (demonstration) projects, often relying on the enthusiasm of single persons or institutes: the early adopters.

It takes another step to incorporate the use of PV into the general policies and regulations of all major players. There is still a lack of strategies aiming at sustainably rooting the application of PV in general policy and day-to-day practice.

2.5   MARKET ASSESSMENT

As is shown in Box 2.1, a coherent (national) policy for market introduction accelerates the introduction of SAPV. This can only be successful when based on a thorough market assessment. Such an assessment provides a powerful strategic view and identifies a limited set of markets where the major part of the turnover is expected. In this respect a Dutch initiative for a strategic approach has proved to be successful in stimulating coherence and joint efforts in the policies of the major market players.

Market assessments outline groups of investors and authorities that may support the structural introduction of SAPV energy services, rather than addressing diffuse groups of end-users or small industries. It is possible to identify larger entities like utilities, conventional industries or government organisations that can be convinced to integrate SAPV in their businesses. In particular, innovative industries are looking for enhancement of their markets, and PV fits perfectly with their search for more sustainable products and services.

This approach gives an overview of all players in a certain Product–Market Combination (PMC). It thus helps to select those PMCs which are most promising, and, in the case of more complicated PMCs, it indicates where problems can be expected. It has for instance proved very difficult to successfully address diffuse groups of end-users or small industries.

A detailed process of market assessment in the case of rural electrification with PV is described in section 6.1.

2.6   MARKET DEVELOPMENT

The challenge of market development is to evolve a technically sound system into a commercial product, including the infrastructure for all the supporting services needed.

The most prominent tools in this respect are full-scale demonstration projects, which allow:

- the early adopters to gain confidence in the innovation
- all services to be developed and field-tested.

The choice of clients is very important, as they should preferably be role models for their sector, for instance a professional PV-powered ferry has a large impact on the recreational boating market, while a small PV-powered dinghy has virtually no influence.

In this phase, the institutional set-up of a range of services such as retail, financing, installation, maintenance, guarantees etc. (as described in Chapter 5) is developed and tested. Sponsorship of these initial activities by governments or large donor organisations is essential.

2.7   OPPORTUNITIES AND LESSONS LEARNED

The most important lesson learned is that the generally assumed multi-gigawatt potential in SAPV has been confirmed and some pieces of that huge market have been filled in. This potential is expected to generate many opportunities for the SAPV sector, including new areas for RD&D.

Opportunities for the Service Application sub-sector are mainly due to the fact that conventional industry is willing to invest in integration of SAPV into its products.

3   Economic aspects

Oliver Paish

IT Power, The Warren, Bramshill Road,
Eversley RG27 0PR, United Kingdom

This chapter reviews economic aspects of different energy options in relation to their scale. The concept of life-cycle costing is introduced, and its importance for SAPV and limitations are discussed.

3.1   OVERVIEW OF EQUIPMENT COSTS

This section provides an overview of the typical costs of different types of stand-alone energy equipment. It must be noted that equipment prices, in what is still a small and immature market, can be highly variable, depending on the country and location, the source and quality of equipment and the taxes or subsidies applied. Table 3.1 therefore provides only indicative costs for a limited range of 'typical' hardware.

Table 3.1: Indicative costs of renewable energy systems

Equipment	Size range	Cost Guidelines
PV modules	50 Wp or greater	$4–7/Wp
PV lighting systems	50 Wp or greater	$8–20/Wp
	<50 Wp	$15–30/Wp
PV water–pumping systems	100–1 000 Wp	$10–15/Wp
Micro-hydro: electro-mechanical equipment	30–300 kW at heads >50 m	$400–600/kW
	30–300 kW at heads 10–50 m	$600–1 200/kW
Micro-hydro: complete installations	30–300 kW at heads >50 m	$1 500–2 000/kW
	30–300 kW at heads 10–50 m	$2 000–4 000/kW
Wind–generator and battery unit	0.1 to 10 kW at 10m/s wind-speed	$2–5/W

In general each energy technology has a certain range of average energy demands which it is most cost-effective in supplying [17]. indicates the appropriate size ranges for the most common options normally considered.

There are prospects for future cost reductions as the different markets develop. For example, the price of PV modules has reduced from $20/Wp to around $5/Wp in the period 1980–97 and is anticipated to fall by at least a further 50% by 2005.

The broad picture can be described as follows (see Table 3.2):

- There will always be a market for portable throwaway batteries for the lowest levels of energy demand, e.g.

Table 3.2: Typical economic power ranges for energy systems

Continuous power:	1 W	10 W	100 W	1 kW	10 kW
Primary cells	■				
PV–battery	■■■■■■■				
Wind–battery	■■■■■■■				
Hydro-electric turbine			■■■■■		
Diesel generator with battery			■■■		
Diesel generator				■■■	
Grid extension					■

watches, torches and radios. Good rechargeable batteries with a PV battery-charger will prove cheaper than disposable batteries over a few years.

- For on–off domestic power supplies for lighting, TV, radio and fan, a battery-based supply charged by PV or wind is likely to be the most cost-effective. If occasional power peaks occur, a back-up generator can be considered.
- For a number of domestic supplies, e.g. for a small village, the economic choice lies between individual battery-based units (solar home systems or commercial battery charging) and a centralised scheme with PV plant, diesel generator or micro-hydro.
- For remote water pumping, the choice is between PV, windpump and diesel pump systems.
- For electrifying larger villages where hundreds of kW are required, the choice is principally between micro-hydro, diesel and extending the grid.

3.2   LIFE-CYCLE COSTING

The initial cost is only one element in the overall economics of a system. Some type of economic assessment is required to determine which system from a number of choices will give the best value for money in the longer run, either for the customer or for the economy as a whole.

The economics of PV and other renewable-energy technologies are rather different to those of conventional small power systems, in that:

- The capital cost of the equipment is relatively high, especially for larger plants.
- The running costs are low and there are no fuel costs.

- The output of the system depends on its location.
- The output of the system depends on the load pattern.
- The reliability is high.

When comparing PV with (for instance) a diesel generating set, the high initial investment costs of the former make PV look unattractive at first sight. However, the picture often changes with an appreciation of the longer-term economic picture. This is normally achieved through the method of *life-cycle costing*.

Life-cycle costing examines all the costs incurred over the lifetime of different systems, and compares them on an equal basis by converting all future costs into today's money. This method is known as a *discounted life-cycle cost analysis:* the result is the *levelised cost*. For instance, a PV array costs more to buy than a diesel generator, but the modules should last over 20 years. The diesel generator might last 10 years, using a certain amount of fuel each year. In this case the analysis period would be 20 years. In addition to the capital cost, the cost of a replacement diesel after 10 years, plus 20 years' worth of fuel, is also included in the cost of the diesel option. The costs of maintenance and repair for the two systems over the whole 20-year cycle must also be added. Depending on the exact figures, either the PV or the diesel system will work out cheaper overall.

Life-cycle costing enables one to appreciate how the various costs involved over a period of time can be simplified into a fixed annual or monthly cost.

3.3 ECONOMIC COMPARISONS

Different energy technologies prove to be more cost-effective in different situations. In general, it is the average amount of energy required per day that dictates the choice of system. In the examples below, life-cycle costing has been used to compare the *levelised* cost of energy supplied by different technologies over their lifetime, as a function of the daily energy required.

PV versus diesel

Figure 3.1 illustrates how the levelised electricity costs of PV and diesel options vary as the daily energy load increases. The effect of variations in both module price and fuel price are shown. PV is almost totally insensitive to the magnitude of load because it is a modular system: a PV system can be sized to meet any magnitude of load without wasting excess capacity. Conversely a 5 kW diesel generator is poorly suited to small loads.

PV versus grid extension

Figure 3.2 compares the approximate levelised electricity costs for extending an 11 kV grid to electrify a small village, compared with providing small PV installations for each household. The conclusion is that, for small loads, it is not worth extending the grid by more than a few kilometres, whereas PV is more cost-effective at small loads, but much more expensive at higher loads. The precise crossover point between PV and grid extension is highly dependent on local circumstances.

Opportunities for cost reduction

Opportunities exist to reduce the cost of PV electricity in the near future:

- Recent trends indicate a steady decline in PV technology costs, especially module costs and selected components

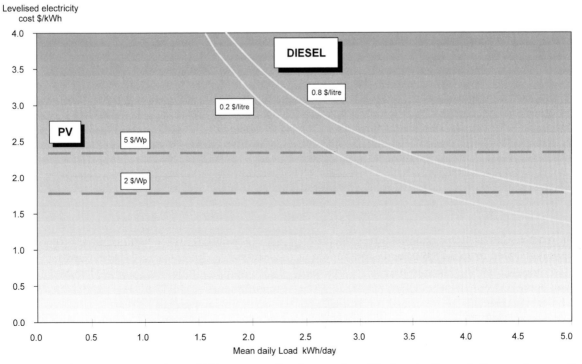

Figure 3.1: Economic comparison of PV versus a diesel generator   (Real costs, 5-year battery replacement).

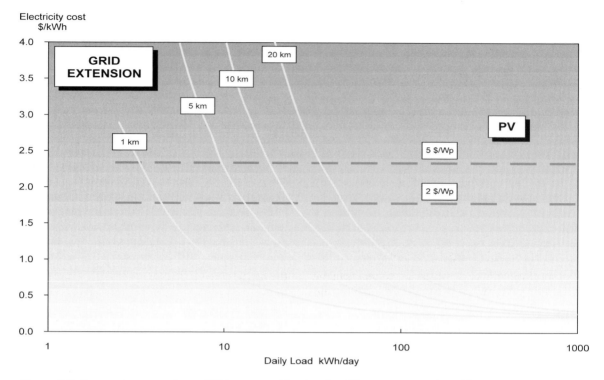

Figure 3.2: Economic comparison of PV versus grid extension (Real costs, overhead line, 20-year life-cycle).

linked to PV systems on the international market. In addition there will be further improvements in component efficiencies.

- Savings are possible through the judicious use of locally made and/or locally assembled components (provided they are of satisfactory quality), as well as the economies of scale in procurement, sales and servicing that enlarged customer bases can provide.

- A strong PV industry together with appropriate policy incentives increases the likelihood of attracting low-cost capital and lowering duty and taxes (or even tax exemption) for the expansion of sustainable new energy models.

Concerning the last point, governments should rationalise duty and tax structures, where these discriminate against PV or rural electrification programmes. Relatively high import duties and other taxes, particularly on PV modules, can severely limit the potential for commercially viable, market-driven programmes by raising the financial costs of power systems.

These cost reductions cannot eliminate the need for adequate financing systems geared to low and middle-income households. Unless such arrangements are in place, PV cannot play a significant role in rural electrification, particularly in developing countries.

The need to overcome the first cost barrier is a *sine qua non* in any country context.

3.4   OPERATION AND MAINTENANCE COSTS

While the design and dependability of most basic PV systems, and especially solar home systems (SHS), allow

a single technician to service a large number of customers, the need for local technical support remains. Users can carry out simple maintenance functions and be trained to assure the good performance of the system. However, field experience shows that very few households can service their system themselves over a longer period of time.

Stand-alone systems are typically used in sparsely populated areas, serviced most effectively by a local representative who can attend problems in a matter of hours or days, rather than the weeks that might be required if the service were provided from a central location.

The number of technicians required in a service territory depends on the number of systems in use, their quality and their accessibility (remoteness, road conditions and available transportation means). A rule of thumb is that no system should be further than 50 km from the service centre, and that a well-equipped technician can service no more than 400 systems. The cost of operating such programmes will be higher where the users are widely dispersed than where they are more concentrated (e.g. higher transport costs, different means of transport, more staff needed).

In some cases the service centre provides not only operation and maintenance services, but also collects fees and administers the programme. It offers effective support for technicians and helps ensure the long-term sustainability of the programme. The cost of operation & maintenance (O&M) must be included in the price of the system or the service.

3.5   ECONOMICS AND REAL LIFE

Although economic analyses have a key role to play in

selecting energy technologies and planning projects, two further issues affect the overall economic outlook:

- Social, technical and personal factors affect individual purchasing decisions.
- Governments can play a decisive role in distorting the economic situation.

Real-life decisions

It has to be borne in mind that customers do not usually carry out a life-cycle cost analysis before deciding to make a purchase. Even if they do, other influential factors play a major role:

- The convenience and portability of the system.
- The ease and speed with which the system can be installed and running: no trench digging for cabling, no cumbersome connections to the grid, etc.
- Avoidance of local contamination, for instance in vulnerable ecosystems where fuel engines are prohibited for fear of spillage.
- The customer's confidence in the performance and reliability of the technology.
- The ease of maintenance and no need for fuel tanking.
- The demand pattern over the year: some applications match well with the insolation (e.g. solar pleasure boats), others have opposite characteristics (e.g. PV streetlights).
- The up-front cost: even when life-cycle costing proves PV to be the cheapest alternative, the initial investment might be a barrier in some cases.

The customers' perception of these, and other issues, all have a value which is not necessarily quantifiable in strict economic terms. At this stage, there is little experience with valuing such considerations, and it remains difficult to compare like with like.

Financial or economic decision-making

Governments and related institutes such as national utilities can have a particularly strong effect on purchasing decisions because of their ability to either tax or subsidise in a variety of forms. Hence cheap goods in one country can be unaffordable in another despite identical manufacturing costs (see Chapter 5).

This raises the difference between taking a *financial* and an *economic* perspective. A financial analysis is carried out from the point of view of the private investor. Thus, the decision for a household or a company will be based on the real financial cost, i.e. actual market prices, inclusive of local taxes, national taxes, import duties and the cost of borrowing the capital to make the purchase. An economic analysis, however, considers projects from the point of view of the economy as a whole. It therefore looks at costs that exclude taxes and subsidies. An economic assessment is the most general case because it reflects the situation of a 'level playing field' in which the prices of goods reflect their real costs.

However, the economic approach must be interpreted as such, because it can lead to misleading results if mistakenly viewed as a financial analysis, i.e. from the buyer's point of view. For example, the local price of diesel fuel can be very much greater than the imported price because of government taxation and high mark-ups in remote areas. In an economic cost comparison, rural electrification via diesel generators may look relatively attractive as a government project, but not for a private investor who may have to pay more than double the imported fuel price.

In carrying out a life-cycle costing exercise, the methodology is the same whether an economic or a financial analysis is being performed; however, certain parameters will have different values. In general, a financial analysis will need to use actual market prices for the goods, a shorter period of analysis and a higher discount rate. Hence a financial analysis usually leads to a higher annual cost for a particular scheme.

4 Financing Aspects

Bernard Bezençon

Atlantis Solar Systems, Lindenrain 4, CH-3012 Bern, Switzerland

Paul Cowley

IT Power, The Warren, Bramshill Rd., Eversley RG27 0PR, United Kingdom

The core of this chapter is concerned with the different financing mechanisms applied to SAPV, together with their strengths and weaknesses. Other financial aspects, such as the end-user's contributions, operation and maintenance costs and opportunities for cost reduction, are also discussed. Although most of the material is presented in the context of developing countries, the conclusions and methodology have general validity.

PV systems are an effective complement to the centralised grid-based power approach, where grid extension is too costly, especially for sparsely settled and remote areas. In such conditions, fuel-independent modular PV systems can offer the most economical means to provide lighting and power for small appliances. Despite these appealing features, PV does not yet have broad market acceptance and faces significant barriers to widespread diffusion. One of the main obstacles is the initial purchase price, which puts this technology out of reach of many potential users in rural or dispersed areas.

4.1 SUBSIDIES AND GRANTS

Governments and donors have provided grants and subsidies for technology promotion and demonstration, regional development, poverty alleviation, environmental improvement and many other reasons. A judicious use of grants and subsidies can help implement rural electrification programmes. However, to assure sustainable programmes, the PV industry should aim at market driven operation, and government support should focus on establishing an infrastructure and building market places through feasibility studies, project design, planning, promotion, training, monitoring and financing (e.g. as loan guarantees to reduce the capital cost of a project).

In PV-based rural electrification programmes, the use of grants or subsidies by governments or donor agencies is not recommended to cover operating costs, as this could undermine the long-term sustainability of such a programme.

At the same time, many governments subsidise conventional energy sources, aiming at social or developmental objectives. This creates serious distortions that hinder competitive assessments and subsequent adoption of the most appropriate energy technology based on provision of the least-cost service. It is therefore of crucial importance that a government provides a *level playing field* for different energy options, whereby an economic value should ideally be attributed to environmental aspects.

In short, subsidising should be directed towards the poorest sectors of society rather than towards technology options. Although market pricing is the ultimate goal, the poorest households may still require subsidies in order to buy and maintain their PV systems. To reach the disadvantaged sections of the population, PV systems should receive at least similar financial support to that provided under conventional grid extension, isolated mini-grids or use of kerosene.

4.2 END-USERS' CONTRIBUTION

PV programmes in the market introduction stage should ideally be operated as businesses. They should generate sufficient revenues to recover capital investment, pay for administrative and support services, cover payment default and, in the case of profit operations, provide satisfactory returns for investors.

In the past, the fees charged under many donor- and government-sponsored programmes were set at levels comparable with the monthly costs of basic traditional energy consumption (kerosene, batteries, candles, etc.) of low-income households. This was based on the assumption that rural consumers have a very limited capacity to pay. Such PV low fees cannot run a programme sustainably over the long term.

Experiences show that customers are often willing to pay more for high valued services than was previously assumed. The end-user valuation of PV household elec-

---

Box 4.1: Sustainability of PV programmes

To ensure sustainability, PV programmes should:

- Set prices to allow full cost recovery and be cautious with the use of grants and subsidies;
- Select only qualified customers with the willingness and ability to pay (this ability may depend on availability of subsidies);
- Ensure that the end-users' expectations are in line with the energy service capabilities;
- Maintain high product quality and responsive services;
- Adopt simplified administrative procedures;
- Select and retain quality staff and management;
- Establish effective fee collection methods and enforce regulations for disconnecting defaulters.

Box 4.2: End-user perception of PV services

The end-user perception of household PV is high. SHS (solar home system) users in Indonesia, Sri Lanka, Senegal. The Philippines and the Dominican Republic indicate that the systems are valued for more than just monetary savings in kerosene or battery costs. Consumer income and expenditure surveys show that a willingness to pay for PV is greater than might be expected from a simple avoided-cost analysis. Rural consumers frequently note the following non-monetary advantages of PV home systems over kerosene lighting and rechargeable batteries :

- Higher-quality light, both in terms of lumen output and colour rendering ability, makes such tasks as reading and studying easier.
- Improved safety levels – SHS eliminate dangers from accidental fires and burns from kerosene and candles.
- There is greater reliability and freedom from fuel need.
- Convenient, instantly available light and access to services (such as TV and radio, without the need to purchase and transport batteries) provide an elevated social status associated with electrification

tricity is high. Consumer income and expenditure surveys show that willingness to pay for PV is greater than might be expected from a simple avoided-cost analysis, as the PV energy services are of a much higher quality than the traditional ones (e.g. PL bulbs compared to kerosene lamps).

4.3   ENERGY SERVICE

The Energy Service is linked to the concept of Demand-Side Management (DSM) or Least-Cost Planning (LCP), already well known in modern utility companies. The idea of an Energy Service was first conceived with the willingness of the local utility to respond positively to consumer needs. In electrification programmes it requires not only maintenance service but also global after-sales service. In PV-based rural electrification programmes the price of the electricity provided by the service would then include the replacement cost of system components (especially batteries), system management information, insurances and so on. It would also promote the use of high-efficiency lights (generally fluorescent) and selected appliances to enhance system performance.

Batteries form a major cost in the maintenance and replacement of PV system components. Customers unable to afford new replacement batteries purchase poor-quality or reconditioned substitutes instead. If these components do not function well, users are likely to become dissatisfied and leave the programme, thereby jeopardising its sustainability. Where consumers' ability to replace batteries is a concern, this cost should be included in the Energy Service's price. This approach offers several advantages: it finances high-quality batteries; it allows for volume discounts from suppliers and it facilitates care of batteries and battery recollection and recycling. In addition, an Energy Service could use the service insurance to cover the costs of vandalism and theft in regions where the risk is too high and no other possibilities exist.

4.4   CREDIT SCHEMES AND FINANCING MECHANISMS

A number of financing constructions have been developed and successfully implemented over time. In this section, four common alternatives are described: (i) payments on cash basis; (ii) consumer financing or credit sales (most widespread); (iii) leasing arrangements; (iv) energy service company (less common). A comparison of their merits is made in Table 4.1, at the end of the section.

Payment on cash basis

Direct cash sales are common in many countries and indeed often the only option. Private vendors receive their products from wholesalers and regional distribution networks. A significant number of systems are sold directly to professional end-users or high-income customers. Cash sales represent the simplest financial vehicle to sell PV. Given the limited disposable income of most rural households, direct cash sales are only adopted by higher-income households and cannot sustain a PV electrification programme for a large proportion of the rural population.

An additional risk with this system is its independence of any installation and maintenance network. In western countries such services may exist and even be linked to the retailer, but in many developing countries these structures are not well organised or not existent. In such cases there is no guarantee of any quality control and deceptions can easily occur, with all the negative consequences for the reputation of SAPV.

Consumer financing or credit sales

In credit sales or consumer financing schemes, banks and dealers provide short-term (one year) to mid-term (three years) financing at market rates to help consumers finance PV systems. This is in many parts of the world a common mode for increasing the sales of durable consumer goods, such as sewing machines, motorcycles, televisions and refrigerators. Users can obtain maintenance services through an annual service contract or on an as-needed basis, especially during the period of the contract. The dealers, who are part of the community, are familiar with their customers' credit-worthiness, and accept the PV system as security. Unlike banks, they do not require stringent security guarantees. Dealers have knowledge of and confidence in their products.

Leasing or hire-purchase arrangements

In leasing or hire-purchase arrangements an intermediary retains ownership of the PV systems until the customers pay for them over a period of time. The PV system is used to secure the lease agreement. If the down payment is too high, it could restrict the potential consumer base that may pay for it.

The intermediary can be a financial institution, private company, cooperative or non-governmental organisation (NGO). It serves as the manager and guarantor of the funds. In many cases, the same organisation also assumes related tasks: it can then register qualified participants, purchase in bulk, provide installation and maintenance services, stock spares, train consumers and perform other administrative tasks.

A different target group for leasing constructions consists of professional clients who are interested in a PV service, but still perceive considerable risks. For instance, municipalities would at this moment not buy PV streetlights, but are interested in leasing them. The limited financial risk that the end-user runs helps him take the step towards the unknown PV technology.

Energy Service Company

Energy Service Company (ESCO) models allow for the most affordable payment schemes, and can thus reach a larger customer base than any other scheme. It brings a least-cost rural energy service with a utility pricing. The ESCO model has several advantages:

- First, spreading the price of PV stand-alone systems over a period comparable to its physical life (ten years or more) can reduce the monthly cost to the consumer.
- The smaller monthly payment makes the system more affordable, allows the ESCO to serve a larger population within its territory and creates a 'critical mass' of demand.
- With a large consumer base, the ESCO can increase system reliability, obtain economies of scale in procurement and in delivery of support services, ease product standardisation and quality assurance, facilitate battery recollection and recycling and help to fight vandalism and theft.

The ESCO sells the energy service but retains ownership of the system; the hardware is neither sold nor leased. The ESCO is also responsible for financial management and administration. While this model is an attractive concept, its long-term effectiveness requires business management skills and technical capacities that may be limited in rural areas.

4.5   INCREASING CREDIT AVAILABILITY FOR PV

The limited credit availability or stringent loan terms to PV enterprises and purchasers constrict the market for PV systems. Some of the reasons for this are unfamiliarity of the lenders with the technology, high transaction costs relative to the size of loans, inadequate collateral, borrowers with no credit history and limited or lumpy cash flows. Some approaches to increase credit for PV are:

- *Seed Capital Fund*: Funds provided by philanthropic organisations or development aid agencies to create a revolving fund used to purchase PV systems. This approach is often used in the early stages of programmes. Examples include: Enersol NGO in the

Table 4.1: Comparison between the main financing credits. Based on: World Bank, 1996 [10]

Financing characteristics	ESCO	Leasing	Consumer Credit	Cash sales
Affordability	High	Moderate	Low	Low
Interest rate	Low	Medium	High	–
Repayment period	Medium	Medium	Short	–
Down payment / Connection fee	Low	Moderate	High	(Full cost at purchase)
Risk to lender	Low	Moderate	High	–
Risk to user	Low	Moderate	High	High
System ownership	ESCO	User (at end of lease)	User	User
Service quality	High	Medium	Medium	Low

Dominican Republic, Solanka NGO in Sri Lanka and the Banpres project in Indonesia.
- *Equity Investments and Debt Financing* by the government: As with grid-based rural electrification, the government finances the initial capital equipment of projects through an equity contribution or loan. This approach is used in Mexico.
- *Asset-Based Lending*: A company could obtain a loan by mortgaging its assets. Unfortunately, many enterprises have limited assets and therefore the amount they can borrow is restricted. In lieu of fixed assets a bank could ask for other forms of security (e.g. post-dated cheques, personal guarantees, bank guarantees). A bank may accept PV systems as partial collateral if they are not too difficult to repossess.
- *Non- or Limited-Resource Financing*: A lender agrees to financing credit, based primarily on the project cash flows. At present, this option is rarely used as most companies do not have sufficient operations experience, although the REC programme in The Philippines did receive such financing from the National Electrification Administration.
- *Green Funds* are an increasingly popular financing instrument as governments allow financial benefits such as reduced taxes.
- *Venture capital* is supplied in the form of equity to young, innovative or growth companies during the development phase (start-up and expansion). In contrast to loan finance, investors play an active role in setting up and developing the company. They aim at high value creation to the benefit of the company and hence also of their investment. Their commitment is limited in duration, generally up to the point of an Initial Public Offering (IPO).

A strong and pluriform PV industry leads to lower risks, and hence increases the likelihood of attracting low-cost capital. However, these cost reductions cannot eliminate the need for adequate financing systems geared to low- and middle-income households. Unless such arrangements are in place, PV cannot play a significant role in rural electrification, particularly in developing countries.

The need to overcome the first-cost barrier is a *sine qua non* in any country context.

5 Institutional aspects

Patrick Jourde

GENEC, Cen Cadarache bat 351, 13108 Saint-Paul-Les-Durance, France

Bernard van Hemert

ECOFYS, P.O. Box 8408, NL-3503 RK Utrecht, The Netherlands

At first glance, most SAPV systems look similar to any other product: produced by a few companies, sold and installed by a supplier and bought by a single client. However, a wide range of institutional aspects influence the conditions and environment in which a new technology competes with traditional energy supply options.

In this chapter, some central issues are briefly discussed, and recommendations on the role of key actors are given.

5.1 KEY ACTORS AND THEIR ROLES

A wide variety of actors can participate: international donor agencies, supra-national, national and local governments, utilities, banks, producers, consulting firms, NGOs, individual and professional clients. However, which role they play and under which conditions varies enormously from country to country and from application to application. It is therefore impossible to give a general picture, and we limit ourselves to a few illustrations.

Industry

The main challenge for the industrial partners is linking traditional industry with the PV industry. The latter should introduce PV innovations that should be incorporated in the products of traditional industry. This way, the sales of PV applications can benefit from the mainstream marketing, financing and distribution services.

International donor agencies

International donors and financing agencies like the World Bank, UNDP and UNESCO, but also smaller agencies, play a major role in promoting PV in general and SHS in particular. Their assistance is most effective when integrated into a broader energy policy or rural development plan. Donors should focus on investment financing and technology transfer. Full-scale demonstration projects should only be implemented if they contribute to market development.

International, national and local governments

Governments should ensure a transparent, supportive institutional and regulatory framework to create a conducive environment for PV. They should support programmes aiming at research, development and demonstration, and develop market enablement strategies. The EC is a good example of a promotor of renewable energies, using a mix of strategies. Outstanding tools therefore will be mentioned in the following sections. Additionally, they can co-finance programmes and act as clients.

Utilities

Utilities play a special role in opening up the markets for SAPV. They may choose to be either end-users or facilitators of SAPV or both. First, SAPV has proven to be an option for rural electrification, a core business for many utilities. Secondly, utilities have many options to facilitate the use of PV [18]. To name a few, it has been shown that:

- the involvement of utilities is a very important factor in supplying credibility to SAPV. The end-users' willingness to invest in SAPV is enhanced when utilities simply associate their name with the applications.
- professional end-users are willing to invest when the utility guarantees the energy service performance, in the ultimate case by offering grid extension in case SAPV fails.
- utilities may themselves successfully develop and market a range of SAPV energy services on a cost-effective basis, through remote pricing. These services can even be applied outside their traditional outlet area, since no grid is involved.

An interesting category is the small local utilities that exist in some countries. These companies buy electricity from the regular utility, and resell it to the end-users. Examples exist where such local utilties offer SAPV services to their clients who live too far for a grid connection.

NGOs

NGOs' main task lies in the implementation of PV programmes and in the provision of after-sales services such as the administration of fee collection and loan schemes and the provision of maintenance services. Proper

---

Box 5.1: SHS for Santa Cruz department, Bolivia

This comprehensive programme aims at providing 10 000 SHSs to a dispersed rural population with limited resources. It appears impossible to realise a programme of such a scale without the involvement of a wide range of institutions. On the positive side, this creates commitment on a broad basis; on the negative side, this hampers the flexibility of the programme.

The producer Shell Solar Energy (Netherlands), through services of the consulting firm BTG, initiated the programme. The Bolivian Partner and client is the Cooperativa Rural de Energía (CRE), a utility on cooperative basis with over 150 000 members.

Financing
The Netherlands-based Solar Investment Fund (SIF) provides a commercial loan for 40% of the US $ 7.5 million transaction, while the Dutch Government under the MILIEV fund subsidises the remaining 60%. One of the MILIEV conditions is that no import duties be raised. As this would violate Bolivian law, the departmental government of Santa Cruz assumes these taxes. The systems remain the property of the CRE: the end-users pay a monthly energy fee of US $ 7.5 for full energy services, including depreciation and component replacements. The end-user also pays for the outdoor and indoor installation costs. As these costs are too high a threshold for many families, the benefiting municipalities provide a soft loan to the end-users through the CRE.

Program modalities
Shell Solar Energy sells installed systems with a one-year system warranty. Therefore, its local representative ALKE has set up an installation network.

Collection of quarterly fees is contracted out to a number of local savings and credit cooperatives. System servicing is provided by a network of field technicians, set up by CRE, who take care for regular maintenance and component replacements.

---

coordination with local and national authorities strongly increases the efficacy of NGO involvement.

A promising role is reserved for an interesting category of NGO: the user associations like SEBA (extensively discussed in Chapter 6).

Joint implementation

Joint implementation is a promising instrument under development, whereby national objectives in the reduction of greenhouse gas emissions may be fulfilled by measures taken in other countries. Joint implementation allows a 'trade' in CO_2 credits between nations, enabling western utilities to invest in renewable-energy projects and gain CO_2 credits. In the case of SHS, PV replaces very inefficient technologies like oil lamps, candles and dry-cell batteries, which cause very high emissions per unit of energy service. Thus, although in absolute figures the contribution to CO_2 reduction is very limited, the costs per avoided tonne of CO_2 are so low that a utility that needs to invest in CO_2 reduction could substantially support SHS dissemination programmes.

5.2   INDUSTRY

Background

Many countries give priority to locally manufactured components or systems and will protect their markets. Some countries, such as Indonesia, have developed a tariff system that levies duties on imported components that are also manufactured locally, but allows exemptions when there is no local production.

Countries such as Argentina, Brazil, India, Indonesia and South Africa produce most components of PV systems, and their production costs are mostly lower than in developed countries.

Local production

Local production should preferably be implemented with the support of large companies through joint ventures. Its feasibility varies by component.

PV modules

Local manufacturing of PV modules is often the first fit, for it is the most visible, valuable and sophisticated component. While the production of silicon wafers is high-tech, the encapsulation of modules is less complicated. Still, a number of important considerations often make it more efficient to buy PV modules on the market:

- Beneficial local manufacturing of modules requires a large internal market (>1 MW).
- The technology is changing quickly. Only large companies can follow such changes and invest as requested.
- The profitability of manufacturing PV modules is not as good as the demand for the product would suggest.

Batteries

Because of their availability, locally made batteries are often used in solar home systems. They are mostly car batteries and their use in PV systems is not optimised. Fortunately, several examples have shown that local production of PV batteries is often a good option:

Box 5.2: Local manufacturing in French Polynesia

A local company, EIG Soler, studied, designed and manufactured in Tahiti between 1980 and 1990 various PV components and DC high-efficiency appliances. This concept was important to reduce the size, and therefore the price, of PV generators and thus increase their competitiveness. The local production included several thousands of deep-cycle or tubular-plate batteries, frames, connection boxes, controllers, freezers and refrigerators, lights, street lights, inverters, colour TV and solar heater panels. 35% of the production was exported to 17 countries.

- The technology and techniques are similar to those for car batteries.
- The same equipment and machinery as for car batteries can be used to manufacture solar batteries.
- Battery production is a modular process: with a few types of containers and plates, a large variety of battery capacities can be covered.

Controllers and lights

These relatively simple electronic devices are widely produced in many developing countries. Although most examples are successful, quality control and international certification remain a matter of concern.

Others

While inverters are high-tech devices not easy to produce locally, other system components like support structures and connection boxes are very suitable. The same holds for DC applications: while DC televisions, pumps and refrigerators may be technologically complicated, other appliances such as street lights and radios are likely to be successfully manufactured locally.

5.3   INTERNATIONAL CERTIFICATION

Optimizing reliability and quality are most important for a new technology like SAPV. Many countries and organisations (including the European CENELEC and the World Bank) have written their own set of specifications, mostly derived from their field experiences. An example is the generally recognised IEC standards for modules. Certification does not only relate to hardware; thus, for example, ADEME and EDF have developed specifications for the use of renewable energies in rural decentralised electrification projects [19]. To harmonize these efforts, an international initiative called PV–Global Approval Programme (PV–GAP) was formed, with a broad-based membership of industrial associations, organisations and individuals promoting PV. However, the standards and testing procedures developed by the GAP do not yet have the status of international standards like those of IEC or ISO [20] .

Special attention is given to the certification of SHS. In many countries, requirements have been defined and implemented nationally, but these requirements are defined for the specific needs of that country, and are difficult to adapt to more general norms. The latest activities in Europe are aiming at developing a range of adequate quality control procedures for systems, components and manufacturing processes, which will foster multidisciplinary perspectives and a collaborative establishment of European industry in developing countries [21].

5.4   LEGISLATION

Some existing practices in taxes, subsidies and regulations work against SAPV, others in favour. A major issue is the widespread subsidy on fuel costs, which puts renewables at a drawback. Fortunately, more and more governments are reversing this situation and developing legal instruments to promote renewables (See also section 3.5). This is done by:

- financial instruments (rationalising import and excise duties, levying eco-taxes, subsidising PV programmes);
- regulations (prohibition of fuel use in ecologically fragile nature reserves);
- convenants (percentage of energy generation sold by a utility must be generated in a renewable way).

It is interesting to mention as an example that the EC is preparing a Directive that lays down rules regarding the access to the electricity market of energy from renewable sources with a view to the completion of the internal European electricity market and in the light of policy objectives to increase the share of renewable energy sources in electricity generation.

5.5   INTEGRATION IN ENERGY PLANNING

In many countries, rural energy planning is still in its childhood. However, integrated energy plans are an indispensable tool for proper coordination when large-scale projects are set up. Local governments, utilities, NGOs and donors usually have completely different expectations about the future electrification of an area.

Under many geographic conditions, SAPV or hybrids will be the only means of electrification, while in other cases PV is used to bridge a few decades before grid extension becomes a reality. Choices in this respect are politically delicate, but need to be taken openly in order to safeguard large investments in PV implementation programmes.

Box 5.3: PV-powered weirs in The Netherlands

Regional water boards own hundreds of weirs to control water levels in smaller and bigger waterways. Several dozens of such weirs, including their telemetric communication system, are powered by SAPV. Fruitful cooperation among a great number of involved parties created the fast introduction of this innovative application.

DG XVII of the European Commission finances a demonstration programme for SAPV applications in Agriculture and water management. The consultancy firm ECOFYS implements this programme in The Netherlands, with additional financial support from the Government through NOVEM. It advises both the water boards as clients and the producers of the various components. Several producers of weirs entered into cooperation agreements with the solar industry to develop and produce PV-powered weirs. The predominant practice is that the water boards own the whole system. However, in a few cases where the water board doubted the reliability of the energy supply, and the utility came in as an ESCO, the reputable utility guarantees uninterrupted energy supply, owns the PV generator and sells the energy service to the water board.

5.6 INSTALLATION, OPERATION AND MAINTENANCE

PV has a maintenance-free image. Although actual maintenance costs are indeed relatively low, they are crucial and an efficient after-sales service is one of the keys to success, as was extensively elaborated in section 4.3. The objectives should be not only to maintain and repair the installation, but also to train and support users concerning load management, preventive maintenance and system expansion. Very often, the after-sales service will be related to the fee collection structure.

To provide affordable and reliable services, it seems most effective to hook up with existing support infrastructure. These can be local NGOs, utilities or commercial companies. However, it should be borne in mind that the average technician is not conversant with PV and will need thorough training.

Battery replacement is likely to be the heaviest load in SAPV maintenance. It is very important for the environment that the worn-out batteries are recollected and properly dismantled or recycled. Especially in large and disperse programmes, this task poses both logistic and technical challenges to the service organisation.

6   Social aspects

Xavier Vallvé and Jens Merten

Trama TechnoAmbiental, Ripollés 46, E-08026 Barcelona, Spain

J. Serrasolses

SEBA, Can Santgrau, E-17182 l'Estanyol, Spain

PV technology can be applied to satisfy any electrical load in any remote area. However, this chapter focuses on those applications (rural electrification) where the succcess of the technology will depend not only on its own merits but also on how well this technology has interacted with the people that daily depend on the PV equipment.

The implementation of PV rural electrification affects several million people in remote areas of developed regions and roughly two billion people in the developing world. The most important drive for this technology is the desire of people in rural areas to raise their standards of living, and as such they play a strategic role in any project. Rural electrification is not only a technical issue, and its implementation has a strong social component. Paying due attention to the social factors will lead to success in rural electrification projects.

This chapter has been written using as a reference the experience of a showcase project realised in the region of La Garrotxa, Spain, under the EC thermie programme, SE/084/92 ES. It addresses some social aspects related to the individual PV electrification. This is done by considering the implementation process in chronological order. At the end of the chapter, multi-user systems are addressed and a global overview on the social changes provoked by rural electrification is given [22]. The focus is on individual electrification providing services comparable to that of the grid, be it through individual electrification or through multi-user systems. It does not consider those minimal systems which only supply the most basic energy needs, such as solar home systems.

6.1   USER INVOLVEMENT DURING PLANNING

Identification of the rural electricity demand

Electrification is an important condition for the economic development of rural regions. Rural authorities are aware of this problem and finance studies for a possible solution. These studies involve both the potential end user and local authorities. Not only the number of sites with deficient electrification are assessed, but also the characteristics of the user:

- Is the user also the landlord, or does he rent the real estate?
- What are the user's energy needs?

- Is the user continuously present on site? If not which are the expected periods of use of the PV system (weekends, summer months, etc.)?
- What are the current activities on site? Which would be the activities in the case of successful electrification?

The initial surveys indicate a need for improved electricity supply for the following reasons:

- difficulties in performing normal daily activities (washing, water transportation, short daylight periods in winter);
- no prospect of continuity for residence;
- no interest in building rehabilitation;
- difficulty in finding tenants;
- obstructions in developing new economic activities;
- high operating costs of inefficient energy systems (fuel for the generator, external charging of batteries, degraded batteries, etc.);
- lack of communication (wireless telephone requires energy).

Public information meetings with potential users

Once the electricity demand has been identified, a programme for the electrification of the region is designed. Public funding is sought from different sources and a financing scheme for the end-user is developed.

The possibility of electrification with SAPV systems is disseminated through newspapers, local radio stations and also individual mailings to the potential users identi-

---

Box 6.1: Rural electrification in the Garrotxa region

The Garrotxa region in the north-east of Spain has an area of 735 km² with 46 000 inhabitants. An initial survey of the region in 1991 detected 220 houses not connected to the electric grid. Of these houses, 70 were permanently occupied, 19 discontinuously occupied, 48 secondary residences and 83 not inhabited. Only 35% of the houses were used by their owners, whereas the remaining 66% were occupied under renting or farming tenant contracts. 59% of the permanently inhabited houses had an agricultural activity and 7% services; the remaining 34% are residences. 38% of the sites already used fuel generators, another 38% did not have any electricity available, whereas the remaining 24% already used a basic PV system.

fied in the survey. Village councils of the rural districts then organise public meetings, where the possible users are informed about electrification by means of PV systems: How do PV systems work, what are the limits, what are the possibilities of PV systems, which electrical appliances can be connected to a PV system, what are the costs and which funds can be made available? During these meetings, a first contact between the interested user and the promoter is established.

It is very important to focus on the need for energy-efficient appliances, but many users do not understand why they should pay more for such appliances just to save 'some' energy. Of course, heating appliances have to be driven by a more efficient energy source like gas or solar thermal energy. Users who do not understand these mechanisms will not be satisfied by a PV system and will state that 'PV doesn't work'.

Barriers

The social situation on site may impede the participation in rural electrification programmes. Examples of such situations are:

- user's economic inability to come up with the payments required;
- lack of interest from landlords in investing in the electrification of a house occupied by farming tenants;
- lack of interest from farming tenants in investing in the electrification of a house that is not their own;
- limited prospects for the user to be able to carry on with the activities on site.

In spite of available public funding, such situations inhibit a significant number of users from entering into the electrification programme.

Solutions

In a number of cases, though not all, solutions could be found by:

- reducing system cost by reducing the power installed, but leaving open the possibility of a subsequent system extension;
- offering attractive payment schemes to the user;
- direct service contracts with the farming tenants, assuring the partial return of their investments if they should leave the site.

On the other hand, during the promotion and dissemination process positive reactions could also be observed:

- several users took a decision in favour of a PV system in spite of not being frequently present on site;
- several farming tenants had a PV system installed without financial contributions from the landlord;
- even when their financial contribution has been sub-

stantial, the users have accepted that the users' association (see Box 6.3) remains the owner of the system.

Evaluation of the individual user's energy needs

Once the user declares his interest in a PV system, the local site is visited and the energy needs are evaluated. It is important to determine not only the energy needs of the contacting person, but also those of other persons on site, for example other members of the family. Insufficient evaluation of the energy needs on site will lead to undersized PV systems and to a low user satisfaction. The points of evaluation are:

- number of persons living on site;
- period of presence (permanent, only weekends, only summer months);
- activities on site;
- electric equipment currently in use;
- possible enlargement of electricity demand.

The factors of this list, although crucial for the correct design of the system size, often cannot be exactly determined.

The decision whether to install an AC or a DC system is not only a technical, but also a social issue. While in principle it is better to avoid the DC/AC conversion, energy-efficient lamps working on DC have often shown poor reliability (lower than that of AC-devices) and are then replaced with inefficient incandescent bulbs – in fact, inefficient halogen spotlights are very fashionable. The poor availability of other DC appliances is another important consideration.

Efficient load management requires user interaction with the system, i.e. a washing machine or deep freezer should not be operated during the night when daytime operation is sufficient. Together with the user, it has to be decided which loads are of low priority. An automated load manager disconnects such loads when necessary.

Again, as in the meetings, the importance of energy-efficient appliances should be underlined. Within electrification projects, assistance to the user in acquisition or assessment of energy-efficient appliances can be provided by a PV users' association.

6.2   USER INVOLVEMENT DURING IMPLEMENTATION

The offer for the PV system

The information about the user and his or her energy requirements is used to calculate the system size and prepare a cost estimate for a stand-alone PV system. Sometimes the system size has to be reduced because of the limited financial capabilities of the user. Financing schemes and available funding have an important impact on the possible system size. It has also to be defined who pays for the system, the landlord or the tenant, or both.

The contract

It is essential to define the responsibilities of the system supplier and the payments of the user in a written contract.

User training

User training is a very important issue from the very beginning of the implementation procedure of a PV system. Users not conscious of the energy demands of the appliances are unable to understand the limitations of a PV system. Such users are continuously discontent with the service provided by the system. The user has to understand that energy-efficient electrical appliances are crucial for the system performance. User training should involve several steps:

- Users manual for the PV system, including the electronic control unit. This manual contains a short text and sketches illustrating the precautions to be taken for proper system operation. In Figure 6.2, an example illustrates the need for energy-efficient appliances.
- Tutorials.
- Information newsletter of the PV users' association, if one exists.

In spite of this standardised user training, most users need individual consultative service. One visit is rarely sufficient to make the user understand the key issues. This makes the consultative service time-consuming and expensive, but it is indispensable for satisfactory system operation. This service can be partially provided by phone, but this is often not available in remote areas.

Placement of the system components

Once the user has accepted the offer and decided to have a PV system installed, the interactive process with the user continues in deciding where to place the different system components.

It is very important that the user is satisfied with the location of the PV array; if not, costly conflicts are created. For example, one user did not want the modules

Figure 6.1: User training is crucial for successful operation of SAPV systems. (Photo: Trama Tecnoambiental).

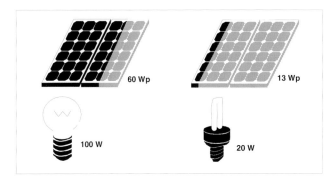

Figure 6.2: Visual illustration of the need for energy-efficient appliances. Diagram: Trama Tecnoambiental.

on his roof. He was finally 'convinced' by the system engineer to agree to this solution, but later on claimed that the PV system affected the water-tightness of his roof. Finally, the placement of the modules had to be changed. Each user has his preferences where to place the PV modules, for example far away from the house or nearby, visible or not.

Another reason why the user should participate in choosing the placement of the modules is that he knows best the special conditions of the site, which may be

Box 6.2: User contracts

Within electrification projects in the Garrotxa region, users sign a service contract with the PV user association as system supplier. Such a contract includes:

User commitments:
- Payments and when to be made,
- Minimal maintenance tasks to be performed by the user,
- Obligation to realise the indoor electrical installation.

Supplier commitments:
- Installation and maintenance of the system,
- Service in the case of system break down,
- Guarantee of operation,
- Insurance of the system.

Figure 6.3: While most users want the panels to be roof-mounted, this owner preferred to place them far away from the house (Can Tarradelles, Spain). Photo: Trama Tecnoambiental.

crucial for proper operation of the system. For example, if the modules on one site were placed too close to the ground, the hens would access the modules producing shadows and leaving droppings on the surface. To be able to give such relevant information, the user has to understand the working principle of the PV system.

Local authorities may influence the decision on the placement of the system for aesthetic reasons, for example in natural parks or on buildings with historic character. In some cases, the landlord has to agree to the installation of the PV system on his real estate. PV module placement should also take into account the risk of the modules being stolen.

Similar considerations concern the installation of the other system components, e.g. the electronic control unit should be easily reached by the user, and the batteries should have their own reserved space with sufficient ventilation.

Installation of the PV system

Involvement of local technicians in installation and simple maintenance leads to an improved relationship between the user and the maintenance service, and as a spin-off creates increased technology transfer and economic activity in the area.

6.3 USER INVOLVEMENT DURING OPERATION

The user interface with the PV system

Efficient load management by the user requires information on the batteries' state of charge. The user has to know when his batteries are empty in order to reduce the loads connected. For this purpose, simple user friendly displays giving a overview of the relevant system parameters have been developed.

Monitoring system

Users often have the firm opinion that they did not use a lot of energy and claim that the system does not to work properly when they use too much energy. Including a monitoring system in the electronic control unit to provide objective information avoids such conflicts. Data monitored is collected at regular intervals by modem connection or direct downloading onto a laptop computer. This data informs the users about trends in their consumption patterns and gives a warning when electricity consumption has sharply increased. The user can then analyse the reasons behind the higher demand and decide whether he or she wants to increase the size of the system or reduce the loads connected.

User satisfaction

Monitoring provides objective information on the use of the PV system, but also allows the system operator to analyse the degree of satisfaction of the user. The key parameters used for this purpose are:

- *Performance ratio (PR)*: Ratio between the energy provided by the system and the maximum possible energy produced by the PV modules. It typically varies from 20% to 70%. Low PR indicates an oversized system providing a reliable service to the user; high PR means intense energy consumption by the user and may indicate low efficiency appliances.
- *Full Battery Ratio (FBR)*: Fraction of days that the battery has been fully charged within one month. High FBR indicates efficient load management. It is interesting to note that the user 'learns' how to use the PV system, which results in an increase in the FBR during operation time, i.e. users tend to use electricity during daytime.
- *PV ratio (PVR)*: percentage of energy supplied by the PV system. Low PVR indicates frequent operation of an auxiliary generator. High PVR means matched energy needs and reduction of the operation time of diesel generators to a minimum.

Based on these three parameters, the users of PV systems can be classified in different typologies based on monitoring data, as listed in Table 6.1.

Users with a high consumption (first two rows in Table 6) need further advice on load management and energy-

Table 6.1: Classification of consumption patterns according to different parameters monitored

PR	FBR	PVR	Consumption	Comments
High	Low	High	High	Undersized PV system but low use of generator, poor load management, user consultancy required
Medium	Medium	Low	High	Undersized PV system, intense use of generator, poor load management, user surely not satisfied with PV system
Medium	High	High	Medium	PV power well matched, low use of generator, good load management
Low	High	High	Low	Oversized PV system, no use of generator, user often absent
Medium	Medium	Medium	Oscillating	High incidence of refrigeration, pumping, rural tourism, seasonal activities, ...
High	High	High	High	Ideal situation making best use of the PV system, user demand perfectly adapted to irradiation, backup generator not needed

efficient appliances; it is likely that these users are not satisfied with their PV system.

6.4 SPECIAL ASPECTS OF MULTI-USER SYSTEMS

The most difficult point in the implementation of multi-user systems for rural villages or islands is to combine the interests of the potential users. For a successful implementation it is important that *all* users are satisfied with the solution found and they *all* support the PV system.

Another issue to be addressed in the case of PV microgrids is the adequate distribution of the limited energy among the users [23]. The fact that several users are connected to the same energy source does not motivate the individual user to buy energy-efficient appliances or handle the scarce electrical energy with adequate responsibility. This problem has led to the failure of several electrification projects in rural villages. The solution requires a distribution and restriction of the electrical energy consumed by each user. This can be achieved by a newly developed electronic device that individually counts, dispenses and limits the energy provided.

6.5 SOCIAL IMPACTS OF RURAL ELECTRIFICATION

On the user level, the satisfaction with PV electrification is high, and the social impact of rural electrification through PV systems is similar to the consequences of grid extension. The most important improvements that have been observed for sites with PV systems are:

- *Quality of life*: One of the aspects mentioned most often by the users is that they pass from rudimentary electricity production systems to a steady AC supply allowing them to have electrical appliances just like

> **Box 6.3: PV user association**
>
> SEBA was born in 1989 after a EU Demonstration project was commissioned in the Solsonès region. Most of the 35 users of that programme decided to create a users' association to share experiences and to find high-efficiency appliances that were not readily available.
>
> When other non-electrified users realised that grid extension was not going to take place, they asked SEBA for help. SEBA was able to manage and commission a project funded by the users and local authorities, thus broadening its initial objectives.
>
> After this successful experience, it was clear that user-driven initiatives were key in developing this market and four more programmes were implemented. By now, SEBA counts over 200 members.
>
> At present, SEBA's main goals are:
>
> - to promote stand-alone rural energy services
> - to seek for public funding to provide this basic infrastructure to low-income groups
> - to supply equipment and services with cost reduction through standardization and bulk purchase
> - to maintain, insure and up-date the energy generation equipment
> - to train users
> - to select and recommend high efficiency appliances.
>
> The positive experience gained in single-user systems in the Garrotxa region has been adapted to multi-user stand-alone systems. Each village creates a local user association integrated as a sub-organisation of SEBA, while maintaining its own institutional structure. This sub-organisation guarantees the operation of the system and collects the fees from the individual users.

people from the city, to have access to information, to ease their daily tasks, etc.
- *Economic activities*: In some cases, availability of electricity has allowed change, increase or diversification of the economic activities. Agro-tourism has been

Figure 6.4: Multi-user system in Escuain, Spain. The users of this system are responsible to keep the trees cut to avoid shading. (The thermal collectors in the front maintain the optimum temperature for the battery room in winter.) Photo: Trama Tecnoambiental.

one of the growing activities in the form of renting rural residences, rooms or houses. Electrification is a must for these activities. Agricultural activities are also supported by PV systems, thanks to the possibility of water pumping (for the animals), electrical fences, refrigerators for vaccines, irrigation system programmers, etc. In some cases it has allowed tele-working. Also some craft activities have been mechanised: e.g. in a rural bakery, where the bread dough was formerly prepared by hand, it is now processed mechanically.

- *Rehabilitation of rural houses*: It was noted, especially in rural villages with centralised PV systems, that many houses in poor condition have been rehabilitated. Previously uninhabited houses become occupied. Some houses formerly without steady use are now occupied at more regular intervals – a first step towards becoming a main residence.

- *Telecommunications*: Rural electrification opens the possibility of connection to the telephone network or wireless phone for remote areas. In one case, the supply of a repeater station for a municipal police radio was made possible using a PV system.

Box 6.4: Social factors affecting rural electrification

The social factors affecting rural electrification through PV systems are illustrated by data from the electrification programme of the Garrotxa region. The table shows the differences between the non-electrified sites (baseline) and those where it was decided to purchase a PV system. From the 220 sites formerly without electricity, 70 now have PV systems. Those sites which have been electrified show:

- *A higher occupation period.* Obviously, more occupation signifies more need for electricity, but also the availability of electricity leads to more occupation.
- *Less farming, but more services or simple residence.* Reasons for this may be that there is more interest in PV systems among those users residing on site, or that there are many sites which have been prepared for permanent residence with a PV system.
- *Land ownership by the user.* It is not surprising that owners are more interested in investing in their houses than tenants are.

	Initial census (% of 220 sites)	Realised projects (% of 70 sites)
Permanent use	51.1	77.1
Farming	59	30
Residence	34	58.5
Services	7	11.4
Property	34	65.7

7 Technological aspects

Sylvain Martel

CANMET Energy Diversification Research Laboratory,

1615 Lionel-Boulet, Varennes, Quebec, J3X 1S6, Canada

Keith Presnell

Northern Territory Centre for Energy Research, Faculty of Technology,

Northern Territory University, Darwin, NT 0909, Australia

From a technical point of view, photovoltaic technology is relatively simple. However, there are still some crucial steps that must be taken at both the design and the installation stages. This section summarises the most important aspects related to stand-alone PV systems. This includes an introduction to PV components and stand-alone systems but also considers issues related to quality, safety and maintenance of systems.

7.1 TYPES OF SYSTEMS

Stand-alone systems are usually categorised into three types, depending on whether they use battery storage and/or auxiliary power source. Figure 7.1 illustrates these three categories [24].

The first type of stand-alone is referred to as *PV-direct* because it powers the load directly, without using any battery. Such a system has the most simple configuration and is normally used either for applications that are not critical and match the availability of sunlight, such as calculators and ventilation fans, or when storage is already part of the system, such as in water pumping. Despite the fact that PV powers the load 'directly', some form of power conditioning may still be required to operate the load properly and maximise the PV output.

The second type of PV system is *PV with battery*. This system includes storage that allows the load to be powered when the PV array cannot supply power directly (e.g. at night and during periods of low sunlight). This is the most common type of PV system as it suits a wide range of applications worldwide.

The third type of PV system, called *PV-hybrid*, includes systems that rely on an auxiliary source to complement the local solar resource, generally a fossil-fuel or wind generator. This type of system generally uses batteries too, for short term variations of sunlight condition (on a daily or weekly basis). It is particularly suitable for applications that are critical (need additional backup) or those found in regions with large variations in sunlight conditions throughout the year, such as high-latitude sites.

System control

All stand-alone PV systems normally require some form of control or power conditioning. The complexity of the control function depends on system user requirements, the type of system and the number of power sources

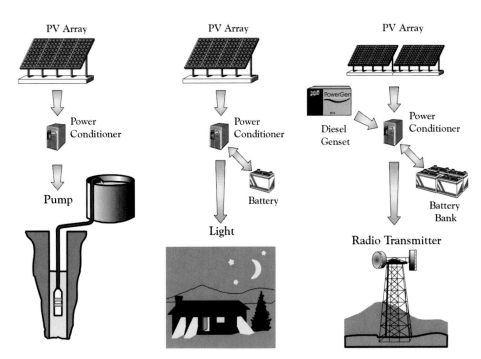

Figure 7.1: Types of stand-alone system: PV-direct, PV–battery and PV-hybrid. Source: *Photovoltaics in Cold Climates* [24].

included. In simple systems, battery-charge controllers interface between the PV array and the battery while the inverter interfaces between the battery and the AC load. However, in the case of hybrid systems, energy management can become more complex in order to improve the efficiency of the system. Some controllers integrate all the necessary functions to run a system.

Nowadays, sophisticated power conditioners offer options such as periodic equalization, energy metering, temperature compensation, multipower source management capability (PV-wind-diesel hybrid for example), monitoring and remote access via modem.

7.2 COMPONENTS

Photovoltaic technology

A photovoltaic cell is a semiconductor device that produces electricity directly from photons (sunlight). A series of cells is interconnected on a panel, with electrical output ranging from 10 to 100 Wp typically. The function of the panel or module is to allow building integration and to protect the cells from the weather. Multiple panels may then be interconnected to form a string, and several strings may be used in parallel to form an array.

Silicon is the main semiconductor used in commercial cells. Panels marketed are mostly made from monocrystalline, polycrystalline or amorphous silicon cells. Many other materials are being developed but have not yet achieved the production level of silicon cells.

While conventional monocrystalline cells have an efficiency of 13 to 16% and polycrystalline about 12 to 14%, relatively high efficiencies (about 18%) are achieved by using new monocrystalline cells with embedded contacts and a grooved surface area. Amorphous silicon is the least efficient of the commercial silicon-based products; while its efficiency is in the 8 to 10% range when new, instability of the material lowers efficiency to a stabilized efficiency of about 3 to 6% after a few months' exposure to sunlight.

The panel output is rated according to an international standard. The unit used is Watt-peak (Wp), which is the panel output under a given light spectrum with an intensity of 1 kW per square meter at a junction temperature of 25°C. In the field, peak power only occurs occasionally, and, as a yearly average, panels will produce no more than 20% of their rated output over a 24-hour period.

Light conditions vary throughout the day and the PV array output will more or less vary accordingly. Among the other factors that affect the PV output, temperature is the most significant. In general, a rise in temperature reduces the performance of the PV array. In a similar way, when temperature drops, the voltage increases and PV panels produce more electricity.

It has been the experience in Australia that in the case of crystalline technology (mono- or poly-), a PV panel frequently operates with surface temperatures above 60°C, when it is only capable of producing around 80% of its rated capacity. However, in the same temperature environment, the efficiency of an amorphous silicon array is relatively unaffected.

The PV array and support structure

A PV array is generally mounted in a fixed position at an appropriate tilt angle (on a building or a separate structure), facing towards the equator. The main advantage of this approach is that it minimises human intervention, but it also limits the performance. Different ways of improving the performance have been tried and these include: manual tilting, tracking arrays and use of concentrators or reflectors. The following briefly describes each of these:

- In stand-alone systems, the array is usually *fixed* to maximize the performance of the PV array during the critical season. This can be the worst period of the year for year-round operation, but could be the summer period, for example for holiday cottages.
- A *manual tilt adjustment* of the array (on a seasonal basis, for instance) introduces flexibility so that output from the array can be maximized during each period.
- The orientation of an array can also be continuously optimized with mechanical devices called *trackers*. These systems can track the sun in different ways, by changing the tilt angle, the azimuth or a combination of the two. The use of a tracker at a very high latitude site in Canada has shown that optimal tracking of the sun more than doubles the output of the PV array during the summer. The main disadvantage of these systems, however, is that they have mechanical moving parts that require maintenance. The reliability is also much lower than that of the PV array itself.
- *Concentrator technologies* include flat mirrors that double the solar incidence, Fresnel lenses, parabolic trough concentrators and paraboloidal dishes. Dish concentrators are able to operate at 300 suns (Solar Systems P/L, Australia), but as the concentration of the different systems increases, so does their dependence on direct radiation. Direct radiation is best associated with a hot dry climate.

The battery

For the applications requiring energy at night or during periods of low sunlight, a storage medium must be used to ensure the autonomy of the system. Most stand-alone systems require storage. The usual storage equipment used with stand-alone PV systems is rechargeable batteries. The following is a brief overview of the different types of battery used with PV systems [25].

Two battery technologies are generally found in PV systems: lead-acid and nickel-cadmium. Both can be found in a variety of sizes and capacity. Nickel-cadmium batteries present some technical advantages over lead-acid and are preferred for some applications. However, they are 3–4 times more expensive per unit of energy stored and consequently lead-acid batteries are more commonly used.

Lead-acid and nickel-cadmium batteries are divided in two categories: *open* units (often referred to as 'vented'), and *sealed* units (also called 'valve-regulated'). When overcharged, batteries produce hydrogen and oxygen, and there is also a consequential loss of water. In open batteries, that loss needs to be made up from time to time. Sealed units, when properly operated, will minimize this loss; for this reason, these are generally considered to be 'maintenance-free' batteries. However, if they are mistreated and overcharged, a valve will let the battery vent, which will result in a permanent loss, since water cannot be added to these units.

Other characteristics, such as the construction of the plate and the type of electrolyte, make some batteries more appropriate under certain operating conditions. For instance, solar-powered telecommunications systems include batteries designed to provide back-up power. Their duty cycle involves infrequent and relatively light discharges compared to batteries used in most other duty cycles. Starter batteries, as applied in vehicles, are designed to accommodate frequent sharp, but shallow discharges. Batteries designed for renewable-energy systems must withstand regular deep discharging. Because batteries are designed to suit a particular duty cycle, it is important that the correct type of battery is selected for a given application.

The power-conditioning equipment

The electrical output of an array is usually modified or regulated in some way to ensure that it meets the requirements of the other components. There are three main power-conditioning devices that are commonly found in stand-alone PV systems: the charge regulator, the power point conditioner and the inverter.

- The *charge regulator* is the most common type of power-conditioning device. It is normally found in all systems using rechargeable batteries. Its primary function is to manage the use of the battery by cutting back current from the PV array when the battery is sufficiently charged. Such devices often include other functions such as disconnecting the load to protect the battery from over-discharging and regular forced equalization to ensure that all cells in the battery are equally charged.
- A *power-point conditioner* (often referred to as the maximum power point tracker or MPPT) is a device that operates the PV array at the voltage that provides the maximum power and then converts this to the output voltage required by the battery or the load. The device allows use of a greater fraction of the energy available from the PV array and may be integrated as a function in the charge controller.
- An *inverter* is required in systems that must supply power to AC loads. This equipment converts DC output from the battery or the array to standard AC

power similar to that supplied by the utilities. Sophisticated inverters often integrate control functions, and bi-directional units are available. The latter can also rectify AC input to produce DC, enabling batteries to be charged from fossil-fuel generators.

7.3   SIZING AND MATCHING OF COMPONENTS

This section does not give complex calculations for sizing or matching components; rather, it highlights the issues that are found to be of particular importance in the process of designing or selecting a PV system.

Energy and power requirement assessment

A first simple, yet crucial, step in the design of a stand-alone PV system is the load assessment. Although straightforward, this step is too often not carried out carefully enough, leading to a suboptimal operation of the PV system. Overestimating the load will ensure a reliable supply of electricity, but the cost of the system will be unnecessarily high. On the other hand, underestimating the load can lead to an unreliable power supply, increased ageing of the batteries and unexpected use of a back-up diesel generator. At worst, the system may fail to supply a critical load. Consequently, all loads must be properly evaluated, both in terms of power and duty cycle (number of hours per day, in a certain period). This indicates the daily energy need.

The maximum-demand assessment is also important, as the system must be sized to have the capacity to power the load. This requires an evaluation of the maximum power that might be required at any time (worst case). Since high values will result in a more expensive system (more storage and possibly a larger inverter would be required), it is wise to consider managing the loads either by reducing demand peaks or by matching demand peaks to renewable energy input peaks.

Forecasting energy production from photovoltaics

The daily average energy production of a PV panel (or array) can easily be evaluated by multiplying its rated output by the average number of 'peak solar hours'. A 'solar hour' is an equivalent measure of one hour of sunlight with at 1000 W/m^2 intensity. The number of 'peak solar hours' can usually be obtained from national radiation data books or can be read on solar radiation maps (see Figure 7.2).

It is necessary to refer to a radiation data book, because the level of radiation is not necessarily a function of proximity to the equator. While any site that is dry and sunny is a prospective site, altitude and degree of atmospheric attenuation are also important factors incorporated in the data books.

However, the energy calculated this way will only be available under ideal conditions. When designing a system, one must account for a number of factors that may

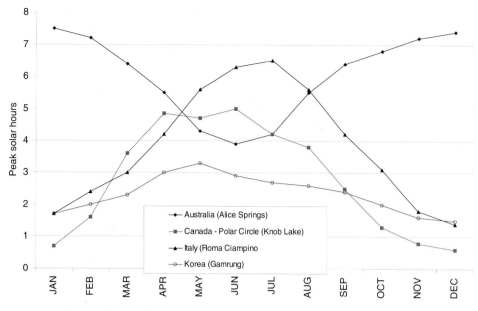

Figure 7.2: Typical solar irradiation curves for some selected sites.

reduce this performance. Such factors may include:

- correction for not operating the PV panel at its maximum power point (panels are rated at this point) if a maximum power point tracker is not used (typically up to 10% losses);
- inefficiencies in storing the energy in the battery and then extracting it (typically up to 20% losses);
- losses of the power-conditioning system for an AC supply (inverter), from 10 to 20% losses;
- the array tilt and position; this is more important for tropical circumstances than temperate areas;
- temperature effects (see section 7.1; up to 20% losses);
- shading, even partial, which greatly affects the performance of a panel;
- dust, snow or ice accumulation, which is strongly dependent on the climatic conditions.

When sizing the system, one should take into account the seasonal variations both in demand for and in availability of energy. For example, a solar pumping system for drinking water will need the same amount of water throughout the whole year, while an irrigation system will have some distinct, short peak periods. On the other hand, the solar energy also varies with the seasons, as illustrated in Figure 7.2. Thus, careful assessment of the critical period is crucial. If power demand curves match well with the availability of solar irradiation, the system can be considerably smaller (and cheaper) than if this is not the case.

Storage – battery sizing

The size of the battery in a stand-alone system (more specifically for a PV–battery system) is directly dependent on four main factors:

- *The autonomy desired.* This is the number of days the battery could provide the loads without any input from the PV array. The longer this period needs to be, the more storage capacity required. This is usually a function of climate and user requirements.
- *The depth of discharge (DOD).* This is how much of the total capacity of the battery is used. The shallower the cycle and the more storage needed to meet demand, the more expensive the battery bank is up-front. On the other hand, too deep cycling of limited batteries reduces their life expectancy, thereby increasing O&M costs. The DOD strongly depends on the application, and can vary from a few percent to as much as 70%.
- *The temperature.* Low temperatures greatly reduce the storage capacity of batteries necessitating additional investment to provide the required capacity. On the other hand, high temperatures are detrimental to battery longevity. The ideal operational temperature is in the 20 to 25°C range.
- *The power demand.* The battery capacity decreases with increasing discharge current. Thus, applications requiring high powers will need larger battery capacity than low-power applications, even if the total energy consumption is the same.

Inverter sizing

The inverter must be able to meet the maximum power demand including the surge produced by loads such as motors. Two ratings are of particular importance for the selection of the inverter: the rating in the continuous mode and the rating in the surge mode (for a very short period of time). To prevent unnecessary losses, special attention should paid to the fact that the inverter is rather inefficient when the load is typically less than 10% of its nominal power rating.

Balancing cost and performance

The prime concern for the user is the cost of electricity generated. Technical considerations, such as the type of cell and its efficiency, are usually irrelevant to users. However, the cost of electricity will vary, depending on:

- the purchase price of the array plus balance of system costs;
- the type of technology;
- the panel efficiency;
- available insolation levels;
- the climatic conditions (including temperature);
- mounting and integration costs;
- the cost of maintenance.

A widely used indicator for selecting a power system is the life-cycle cost as presented in Chapter 3. As several solutions exist for each situation, the life-cycle cost can be evaluated for different options and the choice can then be made. A PV–battery system, for example, can have more batteries and less PV, or more PV and fewer batteries, and still meet the same requirements. Similarly, a PV–diesel hybrid system could use more PV and less diesel fuel over the years, or less PV with the capital saved offset against more diesel and higher maintenance in the future.

In any case there is always a tradeoff between the cost and the performance of a system.

Extensibility

The objective of reducing the initial investment brings up the important issue of modularity and extensibility. For user applications, power consumption normally increases after implementation of the electricity supply, so that most systems are oversized, at least during the first years of their lifetime, and thus their energy consumption is well below the design value. In diesel systems, most of the costs consist of fuel costs, which depend on electricity generation. However, in SAPV and hybrid systems, consumption levels hardly affect the operational cost, which is determined by the fixed costs (depreciation). At present, PV generators are the only modular and extensible component of SAPV or hybrid power supply systems. Incremental steps adding wind capacity, back-up generators or storage batteries are technically complicated and unreliable. Product families with small capacity steps and identical interfaces are rare. System control is tailored for each individual system and is not open to extension [26].

System architecture and control concepts that support modular extension of solar and hybrid systems (cascading) would drastically reduce barriers for first implementation and improve the possibilities for optimization during operation.

7.4 ENERGY MANAGEMENT

Capital is the major cost component of a solar photovoltaic system. Thus, the financial impact of investment in photovoltaics is moderated, or alternatively exacerbated, by the relative efficiency of the appliances or applications that use electrical output from the system. In most cases it is cheaper to invest in improved appliance efficiency than in additional capacity from the solar array.

Applications

Selection of a PV system should be preceded by a study of how the energy is to be used and, where appropriate, consideration of alternatives. Simplistically, electricity is best applied to situations where there is no viable alternative, such as for electronic devices (TVs, computers and facsimile machines), lighting and electric motor drives in a multitude of applications, of which cathodic protection, telecommunications, water pumping, information services and rural electrification are outstanding.

In general, if an application demands much power, PV is not the most economic solution. Examples are large pumping systems and most heating applications. However, direct solar thermal systems are commonly applied for hot water, space heating and drying agricultural produce.

PV refrigeration and cooling systems are commercially available, but these are only economically viable in a limited number of remote applications.

Planning energy consumption

The storage of energy in a battery results in some loss of energy. That is, for every unit of energy taken from a battery, more than one unit will have to be supplied to the battery (commonly between 1.2 and 1.5 units).

It follows that losses can be reduced by limiting the need to draw energy from the battery. This is achieved by matching demand and supply peaks, that is, planning high-energy-use tasks for times when renewable energy production is high, thereby minimizing the need for storage capacity. In the case of a community (mini-grid) system, cheaper daytime rates may be charged in order to stimulate daytime consumption of large loads.

Consideration should also be given to minimizing periods of high energy use. Energy-intensive tasks should be planned so that they do not overlap and cause maximum-demand problems. Excessive current draw from the battery will shorten its life. Also, the capacity of the inverter may be exceeded.

7.5 QUALITY ASPECTS

Chapter 5 stressed the importance of certification as the central approach for quality insurance. While this is a lengthy process, as PV technologies are still under continuous improvement, a need exists for technical quality control in the field. Currently, the quality of some service applications has reached such a level that manufacturers can offer a ten-year system guarantee.

In the pilot phases, PV systems are operated in a research setting, where considerable amounts of technical equipment are used to monitor the performance of all aspects of the systems. Now that many applications have reached a phase of large-scale dissemination, the need for cheap, compact quality-control systems is increasing. In most cases, it is enough to certify the battery voltage and the load current, as these parameters indicate the performance of the system [27]. Of course, proper feedback must be provided not only to the suppliers, but also to the installers, maintainers and even to users.

7.6 SAFETY ASPECTS

In the same way as for quality control, the most important considerations referring to safety aspects were discussed in Chapter 5, where the types of certification were discussed.

The batteries are the most critical components of a PV system in terms of safety. Hazards related to batteries are twofold: acid spillage and the risk of explosion. Spilled battery acid irrevocably destroys any fabric and causes blindness if in contact with the eyes. In addition, an explosive mixture of hydrogen and oxygen is generated in each cell of a lead-acid battery and, if no proper ventilation is ensured, any spark or short-circuit can make the battery explode. Thus, proper precautions must be taken by any person who is in contact with batteries. Adequate manuals for installers, maintainers and users must be provided and followed [28]. Disposal of written-off batteries also entails health risks and must be handled professionally.

7.7 SYSTEM MAINTENANCE

Stand-alone PV systems are almost by definition located in remote areas and thus maintenance requirements must be minimal and predictable. For unattended systems, such as buoys, weirs and telecommunication repeaters, a maintenance schedule of not more than one visit a year is feasible and should be pursued. For attended systems, such as those in remote buildings, a maintenance schedule with weekly, monthly and semi-annual maintenance procedures is reasonable. In all cases, proper planning is required.

Maintenance of the array

Maintenance needs of the array vary according to the climate. In most temperate climates, these needs are zero, but in dusty or snow conditions, regular cleaning can become crucial for proper functioning.

Maintenance of the batteries

Batteries are the components that require most maintenance. For open lead-acid batteries, some means of keeping sufficient electrolyte in the cell is necessary for their optimal operation. For large systems, there are several ways of preventing, or at least minimizing, manual watering. The simplest is to select batteries that incorporate a large water-storage capability as a design feature. Another option is to use recombination plugs or catalytic vents. These simply recombine the hydrogen and oxygen to form water, which is returned to the cell. Automatic watering systems that supplement the cell's design capacity are also an option. They involve installation of a water storage and cell reticulation system and a regulator to maintain a constant water level in each cell.

8 Lessons learned and recommendations

Bernard van Hemert, Geerling Loois

ECOFYS, PO Box 8408, NL-3503 RK Utrecht, The Netherlands

Xavier Vallvé

Trama TecnoAmbiental, Ripollés 46, 08026 Barcelona, Spain

8.1 INTRODUCTION

This chapter summarizes all the main lessons learned and the recommendations emerging from previous chapters and from the showcase projects described in Part II. As such, it is a transition between the two parts of this book. The chapter is structured in line with Part I, while within each section the lessons learned and recommendations are grouped according to the three categories of service applications, remote buildings and island systems.

The main lesson learned is that a generally acknowledged market of several GWp for SAPV applications has been worked out and confirmed for some specific subareas. This implies that a substantial number of remote applications can be powered by a clean and cost-effective technology – PV. A condition is that both industry and governments develop and implement policies to open up the market.

In this regard, six main policy issues emerge in almost every contribution. Although they will be discussed in detail in the following sections, it is worth summarising them in Table 8.1.

8.2 MARKETING ASPECTS

Service applications

Service applications are closest to the mainstream market. Some applications like telecommunications, cathodic protection, signalling and water supply have had a commercial history of more than a decade. And new developments are moving fast: during the last five years several new applications have matured (e.g. parking meters and public lighting in bus shelters).

The marketing approach described in Chapter 2 particularly applies to the area of service applications. It consists of identifying the most important product–market combinations (PMCs) and subsequently targeting limited resources at opening up these markets. Of crucial importance is the cooperation between all major players in that sector: be they industry, utility, local authority or

Table 8.1: Major policy issues regarding the large-scale introduction of SAPV, divided by category

Issue	Service applications	Remote buildings	Island systems (hybrids)
Hardware integration of PV panels, regulators, displays etc. into conventional products is the main technological challenge to mainstream PV-powered appliances.	***	**	*
Integration of SAPV products in mainstream operations of 'conventional' industries allows benefits to be gained from existing, highly competitive production systems and networks for financing, distribution, sales and service.	***	***	**
Cost reduction of the system, not only of the PV panels but also other components and support services, is a *sine qua non* for large-scale commercial expansion of PV.	**	***	**
Financing instruments are required to overcome the initial investment barrier .	**	***	
The lifetime, price, storage capacity and environmental impact of **the batteries** remain major concerns for SAPV applications.	***	***	***
User training in energy-saving practices and load management is a crucial investment to make an application successful.	*	***	**

* Requires little attention at this stage
** Requires attention at this stage
*** Requires priority attention at this stage

others. Emerging efforts and plans seem to justify expectations.

Lessons learned

- One of the main drivers for this development is the growing number of initiatives for collaboration between the PV industry and conventional industry. Economics are always the driving force behind such initiatives. Commercial advantages are expected: e.g. gain of market share through a broader product range or lower maintenance costs while increasing the quality of service.
- Government support in terms of sponsoring market assessments has been shown to accelerate the above-mentioned process of cooperation between market parties.
- For the owner/administrator, SAPV energy services provide a more reliable power supply that displaces high labour costs or expensive fuel.
- The identification of the main PMCs provides fast insight into the potential of SAPV in a limited set of products (both hardware and software) for homogeneous target groups.
- This coherent strategic approach helps convince actors to invest effort and capital in the application.
- Cooperation with conventional industry and service companies is essential to benefit from existing infrastructure and services. In this cooperation, conventional industry brings experience and infrastructure in (high-volume) sales, distribution and service. The PV industry brings innovation, a reliable power supply and a green image.

Recommendations

Despite these positive trends, most SAPV applications are still at a very early phase of market introduction, and the following issues need attention:

- The industry and other market participants should systematically analyse the national and international markets for SAPV to produce specific market introduction plans for a selected number of promising product-market combinations. Financial support from national governments and international support organisations, such as the IEA and the EC, will accelerate this process.
- Market participants should survey and assess the strategies that have been implemented in the various showcase projects to gain the involvement of larger investors in markets of some volume. Cofinancing from national and international authorities could facilitate such surveys.
- On the basis of the results, implementing agencies should further develop these strategies.
- 'Conventional' industries should grasp the opportunity to extend their 'standard' product range with PV-powered products.
- The industry should pursue twinning of PV suppliers with suppliers of related components (e.g. PV-driven motors and boats) as a first step towards integrated product development and marketing.
- All parties involved, including research institutes, should evaluate the development of such cooperation processes in order to learn from their dynamics, to adopt the positive aspects and to validate today's expectations.

Remote buildings

Although the area of remote buildings is by far most important in terms of Watt-peaks, structural analysis of marketing strategies is relatively new. It consists of a diffuse consumer market, an important part of which is demand-driven. Simultaneously, an increasing number of rural electrification programmes are targeting this sector.

Lessons learned

- The market for remote buildings is traditionally a typical consumer market with a broad spectrum of individual end-users, reached through many different channels.
- However, the market is relatively easy to open up, because electrical energy is in high demand and SAPV has many advantages over its major competitor, the diesel generator.

Recommendations

- Project developers should focus on developing large-scale introduction schemes, because there have been sufficient positive experiences with technical, institutional and commercial aspects.
- 'Easy access' and 'service assurance' should be key elements in a marketing strategy for this sector, as these are perceived by the clients to be major problem areas.
- The recommendations regarding the importance of twinning PV suppliers with conventional industries, as described for the service applications, also hold for the remote-building applications, although here the industrial partners are limited to the building industry. Prefabricated buildings and leisure cottages are the most promising areas.

Island systems

'Island systems' is by far the 'youngest' of the three categories in terms of marketing, and little practical experience has been gained.

Recommendations

- Potential clients, such as utilities and local authorities, should be invited to participate in demonstration projects at an early stage, so that they can get acquainted with the technology and start developing solutions for the institutional aspects from the very beginning.
- A first step in wider application of PV-hybrids should be to link up with those situations where existing generators need expansion or upgrading.

- Therefore, governments and international donor agencies should commission a study to survey diesel power supply systems for islands that will need upgrading in the coming years, including a first assessment of the feasibility to apply PV–diesel hybrids.

8.3 ECONOMIC ASPECTS

Chapter 3 introduces the concepts of life-cycle costing and immediately places this calculation method in a broader perspective. The experiences are quite similar for all sectors, and most conclusions and recommendations have general validity.

General lessons learned
- The real lifetime of PV panels (20 years minimum) is much longer than the depreciation period mostly used in economic feasibility cost calculations (often 10 years, based on, for example, the lifetime of a diesel generator). This negatively affects the economic viability of PV.
- It is not only financial considerations that contribute to the decision to apply PV. A whole range of economic criteria exist, although some may be hard to quantify: ease of maintenance, high reliability, independence, flexibility, status, environmental aspects, green or high-tech image, etc. All these can play decisive roles in favour of SAPV, even when solely financial considerations would yield a negative outcome.

General recommendations
- Governments that want to promote renewable-energy options should apply an integral set of price and tax measures as described in section 8.5.
- Cost-calculation methods using the concepts of life-cycle costing, as opposed to a depreciation period, should be generally applied when analysing the feasibility of SAPV systems.
- Economic considerations that are difficult to quantify, such as ease of maintenance, high reliability, independence, flexibility, status, environmental aspects, green or high-tech image, should be taken into account when considering the introduction of SAPV.
- Industry should sustain and increase efforts to reduce the cost price of PV components through mass production and improvements in production technology.
- Policy-makers in sparsely populated areas of developing countries should consider rural electrification through PV power systems, because grid extension is not a feasible option for the predominantly small loads.

Service applications

Lessons learned
- SAPV has become broadly accepted as a reliable and economically competitive solution for many remote service applications with modest energy needs.
- For many low-energy applications (parking meters,

flange lubricators), connection to the nearby grid is more expensive than applying SAPV.

Recommendations
- International bodies such as IEA should further develop and promote the concept of 'price for the service', as opposed to the limited 'price per kWh'.
- Donor agencies should commission a study to survey industrial and consumer low-energy appliances that could be powered by SAPV, and assess the market potential for each of them.

Remote buildings

Lessons learned
- There are now a number of comprehensive comparisons between grid extension and rural electrification through SAPV, and these clearly prove the value of SAPV.
- Based on these assessments, in a new development some utilities have successfully started to offer 'wireless service' instead of grid extension.

Recommendations
- Both the large utilities and the small local electricity-selling companies should consider the economic feasibility of offering 'wireless services' instead of grid extension in remote areas.

Island systems

Lessons learned
- Limited experience has shown that PV–diesel hybrid systems perform well in providing reliable energy at competitive pricing to isolated islands and villages.
- In a newly electrified island or village, the demand for energy shows growth rates that are significantly higher than those commonly used for projections.
- The relatively high costs of remote maintenance are a major factor in the pricing.

Recommendations
- Hybrid systems should generally be considered as a first option, as they prove to be considerably cheaper than sole PV generators, and can mostly compete with diesel generators, depending mainly on the price of fuel.
- Industry should pursue modularity in hybrid system components. This will allow the system to expand easily, so as to keep up with the increase in power demand. This in turn will substantially reduce the initial investment, and thus the cost of energy.
- Industry and project designers should further develop efficient and reliable maintenance systems, in order to reduce costs. Remote monitoring is one of the key issues.

8.4 FINANCING ASPECTS

Chapter 4 analyses the impact of different financial instruments (grants, subsidies, soft loans) that can be used to promote PV applications and stresses the importance of a level playing field for PV *vis-à-vis* other energy options. Four financing systems are described below: cash sales, credit sales, leasing or hire purchase, and energy service companies (ESCO). Most experience has been gained so far with the first two systems, but there are several reasons why the last two should be developed.

General lessons learned
- The initial investment is the main barrier for the application of PV, especially for remote buildings and island systems.
- Utilities that expand their package to include SAPV services provide a major source of finance.

General recommendations
- Governments and donors should focus on developing sound financial instruments to overcome the first investment barrier.
- Industry and donors should finance demonstration projects to further develop leasing and hire-purchase arrangements. For private owners of remote buildings this can act as a powerful financing instrument, while for institutional clients with service applications and island systems it can help overcome reservations.
- Cross-subsidies, as commonly applied for rural grid extension, should also be made available to stand-alone PV schemes that guarantee long-term and sound electricity supply.

Service applications

Lessons learned
- In the market for professional service applications, the traditional financing mechanisms and channels are mostly applied whether SAPV is considered or not.

Recommendations
- Utilities should expand their operations to supply full services (e.g. public lighting) or alternatively energy guarantees to isolated rural SAPV applications.

Remote buildings

Lessons learned
- Cash and simple short-term credit sales allow for fast and easy spread of PV technology, but entail the risk of not building proper after-sales service structures, which are essential for regular maintenance and quality control.

Recommendations
- Financiers should include the investment in the PV system in the mortgage of the building.
- Energy Service Companies should be considered as the most promising institutional set-up for providing energy services to the disadvantaged dispersed rural population, which would otherwise remain deprived of electricity because of their limited financial capacity.

8.5   INSTITUTIONAL ASPECTS

Chapter 5 illustrates, with several examples, the importance of a favourable institutional setting.

General lessons learned
- Government policies have a strong influence on the competitiveness of PV in comparison to alternatives such as diesel and grid extension. Tax rates and subsidies in a variety of forms complicate straightforward comparisons. This can work either way: subsidies can be given to diesel prices and grid extension programmes or to PV systems. The policies of many authorities in subsidising PV products have their limitations, because they do not stimulate economic competitiveness.

General recommendations
- To governments that want to promote PV, the ultimate goal should be an open market. Therefore, they should develop long-term strategies that include:
  - guaranteeing a level playing field by assuring that all energy options have similar access to direct or indirect subsidies;
  - developing tax mechanisms that make a charge on pollution, such as CO_2 emissions, and on exhaustion of natural resources; thus indirectly benefiting renewable sources of energy like PV;
  - financing technology development;
  - financing establishment of infrastructures;
  - building markets by financing feasibility studies, project design, planning, promotion, training and monitoring;
  - financing demonstration programmes;
  - providing loan guarantees to reduce the capital cost of a project;
  - maintaining a cautious attitude towards investment subsidies for PV, while not ruling them out as the ultimate solution.
- Governments and international agencies should push the development and introduction of international standardisation and certification, and build the institutions to implement testing and certification.

Service applications

Lessons learned
- Utilities and ESCOs offering and guaranteeing decentralised energy services for service applications help build confidence among potential clients. In addition, they increasingly provide full services to clients who do not want to get involved in energy supply issues.

Recommendations
- Industry and research organisations should build up experience of large-scale implementation programmes, because, up to now, most implemented projects have been on a small to medium scale.

Remote buildings

Lessons learned

- Up to now, the necessary infrastructure for sales, financing, maintenance services etc. has been set up at project level. To be able to service substantial markets, this has to be generalized and institutionalized.
- In many cases, local authorities and utilities tend to go along with 'business as usual' and throw up barriers against PV systems.
- User associations are well rooted amongst end-users and can play a crucial role in increasing the quality of rural electrification services. By virtue of their organisation, they become strong negotiation partners for suppliers, project designers and local authorities. In addition, they may assume tasks such as training users, providing maintenance services and collecting user fees.

Recommendations

- Demonstration programmes should be widely implemented, for example as a means to gain support from local authorities and other stakeholders. However, the technology applied should be sufficiently developed in order to avoid any problems that will be counterproductive as regards the whole demonstration programme.
- Designers of rural electrification programs should consider facilitating user associations that can develop into influential stakeholders. In addition, these associations can contribute to tasks such as training users, providing maintenance services and collecting user fees.
- When opening new markets or introducing new products, industry should ensure sound service and after-sales systems. Linking to existing traditional industry infrastructures is highly recommended.
- Industry and research organisations should build up experience of large-scale implementation programmes, because, up to now, most implemented projects have been on a small to medium scale.
- In these large-scale implementations, service structures should be institutionalised rather than set up at project level.

Island systems

Hybrid island applications are still in their infancy, and need more full-scale experiments to develop and field-test institutional and financing frameworks before this application can move ahead.

8.6 SOCIAL ASPECTS

Chapter 6 discusses the importance of user involvement in all stages of the introduction of PV in rural electrification schemes. The same applies – albeit to a lesser extent – to other SAPV applications.

General lessons learned

- Provided the system is well designed, and services are readily available, end-users are quite satisfied with SAPV services. Noise reduction and ease of operation and maintenance are among the most appreciated benefits.
- In virtually all stand-alone consumer systems, energy demand has proven to increase considerably with time, especially during the first few years.
- Rural electrification schemes deal with a variety of individual end-users, whose demand patterns (apart from the general increase mentioned above) are fast changing and relatively unpredictable. This complicates system sizing, interferes with systems performance and affects supporting services, such as guarantees, maintenance schemes and user training.
- Visual interactive displays, though still rare and far from fully developed, are a useful tool for proper load and battery management.

General recommendations

- The industry should further develop visual interactive displays for load and battery management, and integrate them into the control system instead of considering it an add-on. Of course, the long-term objective should be a system that simply delivers energy, without any need for care from the end-user.
- The industry should develop smart controllers that 'learn' to adapt settings on the basis of weather conditions, cyclic consumption patterns and so on.

Service applications

Lessons learned

- When considering social aspects, it is useful to make a distinction between service application with and without direct end-user interference (e.g. PV leisure boats *vis-à-vis* public lighting or parking meters).
- For those service applications where there is direct intervention by lay end-users, thorough user training has been shown to be crucial for a successful introduction. A good training programme matches user expectations with the possibilities of the system, promotes rational use of energy, improves on battery management and lowers maintenance costs.

Recommendations

- Where relevant, end-user training should receive due attention (training aspects are further elaborated under 'remote buildings'.)

Remote buildings

Lessons learned

- SAPV applications can structurally improve the living conditions of the user.
- Generally, a high degree of satisfaction exists among SAPV end-users.
- Thorough user training has proven to be crucial for the successful introduction of rural electrification. A good training programme matches user expectations with the possibilities of the system, promotes rational use of

energy, improves on battery management and lowers maintenance costs.

- User involvement in decision-making pays back (e.g. the siting of a PV array to avoid shading, theft, etc.).
- The choice between AC and DC is not straightforward. Arguments are:
  - The user has already AC devices.
  - The availability of DC appliances is very limited and their prices are high.
  - There is a shortage of experienced DC installers, unlike AC ones.
- Superficial interviews with the end-user regarding energy demand yield unreliable predictions, providing either under- or overestimates.

Recommendations

- Programme designers and implementers should invest a considerable amount of time and expertise in thorough user training, because failing to do so hinders successful implementation.
- In view of the variation in local conditions, training programmes and materials need to be redeveloped for any specific application.
- User participation should extend beyond mere training, and include participation in planning and design.
- Retailers, installers and maintenance technicians should also receive training and their involvement in improving system strategies is crucial.
- System designers should carefully analyse energy demand based on practical measurements rather than on questionnaires.
- The industry should develop a wide range of energy-efficient DC appliances, preferably with zero stand-by consumption.

Island systems

Lessons learned

- Energy management in multiuser systems is a matter of attention. Individual metering is expensive but seems unavoidable.
- Users can play a positive role in optimising load patterns, provided they have been made aware of the problem and stimuli exist for them to participate in this way.

Recommendations

- The industry should further develop cheap meters and load-control equipment for multiuser SAPV or hybrid systems.
- Program designers should implement technological, economic and social instruments to promote optimal load balancing over the day and over the week.

8.7 TECHNOLOGICAL ASPECTS

In Chapter 7, an introduction on the technical aspects of SAPV has been given. This chapter, together with the showcases, yield the recommendations given below.

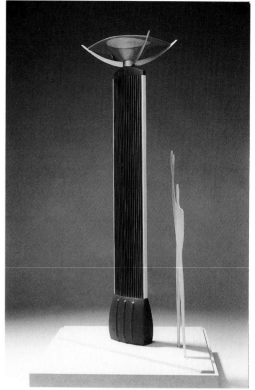

Figure 8.1: Full integration of PV in street lights (model). Photo: ECOFYS.

General recommendations

System integration

Hardware integration seems to be the most important technical development in mainstreaming SAPV in all three areas: it avoids theft problems, reduces costs and improves the appearance of the product. Real integration has only been realised to some extent and in a few cases (boat hulls). The following recommendations can be formulated:

- Industry should boost the development of flexible cells and thin films.
- The industry should improve the efficiency of PV cells, because size reduction is a precondition for optimal integration.
- Product designers should not only consider the modules that are to be integrated into the applications, but also batteries, electronics and control units.

Batteries

Batteries are the most vulnerable and troublesome component of SAPV systems, and the following recommendations are crucial for the further development of SAPV applications:

- Industry should improve the efficiency and lifetime of PV batteries.
- Industry should improve the technology for adequate and easy monitoring of the state of charge (SOC) of batteries.
- This technology should be incorporated in controllers that provide better battery management and protection.

Such improved controllers should also include self-learning features as discussed earlier.

- Authorities and project designers and installers should ensure proper collection, recycling and/or disposal of old batteries to avoid serious contamination of the environment. This point needs special attention in dispersed remote buildings, where many system owners are only loosely connected to a range of different service centres.
- In situations of high DOD, system designers should consider overdimensioning the batteries; although this will increase the initial investment, it will considerably improve the lifetime and may positively affect the life-cycle costing.

General

- Safety aspects in SAPV installations need special attention, not because hazards are greater than those in traditional electrical installations, but because they are different. Most hazards come about as a result of the unfamiliarity of the installer with DC installations and batteries. Thus, training of technical personnel is recommended.
- System designers should consider special conditions, such as cold climates and high salinities, while the industry should adapt specialized products to these conditions.
- Donors should finance research institutes and industries to implement technical demonstration programmes so that field experience with a new application or technology can be gained. It is important to distinguish clearly such field testing from full-scale demonstration projects aimed at market introduction, where the focus is on real-life conditions.

Service applications

The main focus on technical improvement in this area is on hardware integration and the improvement of batteries.

Recommendations

- Product developers should thoroughly analyse the electrical engineering of an existing service applica-

tion, because it is likely that a considerable amount of energy can be saved (e.g. boat hull and propeller design, flange lubricators, reverse-osmosis water treatment).

Remote buildings

The major need for technical improvement in remote buildings is in the fields of user interfaces, improved batteries and improved energy and battery management.

Recommendations

- The industry should pursue standardisation of components, which enables modular building of systems, simplifies design, lowers production costs, reduces installation time and facilitates repair.

Island systems

For island systems, technical improvements focus on questions of control systems, user interfaces, cheap energy meters and modular extensibility.

Recommendations

As has already been discussed in previous sections, technical development of island systems should concentrate on the following technical recommendations:

- The industry should design components so that modularity of systems becomes straightforward, in order to allow easy expansion and reduce the initial investment.
- The industry should further develop remote monitoring systems, because these promise reductions in the maintenance costs.
- System designers should pursue optimal matching between PV and generator, which needs proper engineering.
- Industry should develop cheap individual metering systems.
- Industry should develop better controllers and user interfaces to reduce operational complexity.

Part II

National showcase projects

The aim of Task III was to improve the state-of-the-art of the SAPV of the early 1990s (as illustrated in the examples booklet [2]). The combined efforts of Task III members has succeeded in improving the systems and structures of the 14 national showcase projects, as presented in this section of the book. The showcase projects are grouped according to their category: service application, remote buildings and island systems.

SAPV dual reticulation reverse osmosis water treatment

Australia

APPLICATION

The application of renewables provides reliable power for a unique, dual reticulated water supply (potable/non-potable) and water treatment plant. The overall system consists of a hybrid PV/battery/gas (LPG) system for supply of general power requirements plus stand-alone PV for water treatment and supply. Activities include the construction and commissioning of an entirely new sys-

tem coupled with extensive monitoring of all aspects of operation. The population of the remote, arid-area site varies between 1 and 120 persons. The climate varies from subzero winter temperatures to as much as 50°C in peak summer. Water treatment by reverse osmosis (RO), accompanied by a simple pre-filtration step, removes particles and dissolved salts from the source water such that the final product conforms to the Australian national drinking water guidelines. The system lends itself well to remote applications where a minor maintenance call-out might cost in the order of a thousand dollars.

TYPICAL SYSTEM CONFIGURATION

- SAPV for borehole pump to general (untreated) supply high-level tank and non-potable reticulation network.
- Separate SAPV for RO unit delivering treated water into potable supply tank and reticulation network.
- Blowdown, the concentrated saline solution discharged from the RO unit, is transferred back into the untreated water tank for non-potable use, thus eliminating any water losses.
- Suitable system capacity: up to 1 000 L/day (potable, but no real limit)

PV Bore pump power supply

PV R.O. pump power supply

Cartridge filters

R.O. membranes.

Ground water source

22 KL raw water storage tank

Concentrate to raw storage tank

Permeate to treated water tank

22 KL treated water storage tank

Non-potable reticulated water supply.

Potable reticulated water supply.

The 4 kWp array in the front is part of the hybrid power station. The 480 Wp array on the tank tower powers the RO system.

RO unit with pre-filters (cartridge)

ADVANTAGES

The major drawbacks of remote water treatment are poor power-supply reliability and high system maintenance requirements.

- The advantage here is the application of reliable PV-based power supply with an appropriately sized, simple to operate, but advanced water treatment process with minimal maintenance requirement.
- The RO unit operates at low pressures, which significantly reduces membrane fouling and increases operational life.

MARKET

Market potential exists for small, remote communities especially where no reliable power supply exists and water resources are poor both in quality and quantity, and where maintenance programmes are expensive due to remote access (remote systems in Australasian region).

INNOVATIVE ASPECTS

- Dual reticulation system; partial treatment of water source where full treatment is not warranted and expensive.
- No brine waste disposal required; nil water losses. This is especially important where water resources are scarce.
- Over-dimensioning maximizes reliability.
- Back-up storage and safety systems minimize possibility of plant failure and allow more flexible maintenance response time.

- PV operation at design load all year during all sky conditions.
- Membranes have easily reached the end of their design life and in some cases have not required chemical cleaning.

FINANCIAL ASPECTS

The system is economically feasible for remote applications where no reliable power supply is available and the water source requires treatment. Potential cost reductions are:

- RO operation (membrane type, configuration and operating pressure)
- materials in general
- optimisation of PV cell requirements.

LESSONS LEARNED

Lessons learned from other remote water treatment systems are: failure due to unreliability of power supply and system failure requiring high call-out frequency and technical maintenance requirement. This system has resolved these difficulties and represents a total, reliable answer to remote power supply and water treatment options.

REMAINING ISSUES

- Reduction/optimization of PV cell requirements.
- Optimization of RO configuration, operating pressure and membrane type.
- Alternative treatment evaluation such as ultra filtration, micro-filtration etc.

Hybrid power systems for mountaintop repeater stations

Canada

Photo: NorthwesTel

APPLICATION

With most of its territory sparsely populated, Canada has many applications that need an autonomous power supply. In these cases, the standard power solution usually consists of a diesel-powered generator. However, there is a shift towards PV–diesel–battery hybrid systems that represent a viable solution even in the north.

In this example, a microwave radio repeater is located in the Nahanni Range Mountains, Northwest Territories, at 61.5°N and 2 600 m above sea level. Before the PV upgrade, the station was powered by diesel with a battery bank to reduce run-time and fuel consumption and to increase reliability. By adding a PV array, it is expected that 75% of the electricity needs of this site will be supplied by solar energy, making it unnecessary to run the generators for most of the summer.

TYPICAL SYSTEM CONFIGURATION

The power generation and storage equipment of this station now includes:

- a 1.5 kWp PV array
- two 10 kW diesel generators
- a 1400 Ah 24 V battery bank.

ADVANTAGES

The main advantages of hybrid power systems over a diesel/gasoline generator alone are:

- lower O&M cost
- lower life-cycle cost
- trouble-free operation
- higher reliability
- reduced pollution.

MARKET

In the northern region of Canada the solar resource is abundant in summer, but limited in winter. This makes PV–diesel hybrids generally the best option for many applications.

- In telecommunications only, hundreds of sites can be converted into hybrid systems cost-effectively.

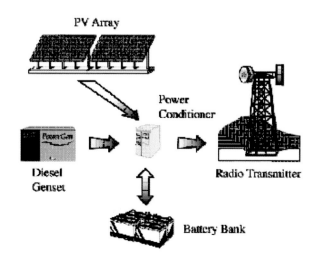

The remoteness of such a site makes remote monitoring essential so that information can be obtained about system performance and so that the system can be controlled. Photo: NorthwesTel.

Manual de-icing of structures is very difficult and usually not cost-effective at mountainous sites. Photo: RCMP.

- The total technical potential in the Northwest Territories exceeds 0.5 MWp.

INNOVATIVE ASPECTS

This project has been:

- A first experience for NorthwesTel with a PV-hybrid system.
- Remotely monitored, to allow evaluation of the PV array performance in harsh climatic conditions.

To date, no problem has been reported with the PV system, making it a successful experience for the utility.

FINANCIAL ASPECTS

While the cost of the PV upgrade at the Nahanni station was about US $35 000, the annual savings are about US $7 000 (of which about $5 500 is related to diesel fuel). This project should pay for itself within five years, leaving 15 more years of savings. A PV upgrade is particularly cost-effective in cases where batteries are already part of the system.

Passive solar de-icing technology for a PV module under evaluation at a rime-prone site. Photo: Global Communication Contractors Inc.

Cost estimation of a 1.5 kWp PV 2 x 10 kW DG system (US $)

Component / element	Present costs
PV modules (1.5 kWp)	8 000
Diesel generators (2 x 10 kW)	24 000
Batteries (34 kWh)	6 000
PV BOS (structure and controls)	4 000
Diesel BOS (tanks)	8 000
Transportation and installation	15 000
System cost	**65 000**

LESSONS LEARNED

- On-site labour must be minimized to reduce costly installation time (man and helicopter). Pre-assembly

of components and structure greatly contributes to this.
- Professional PV services ensure that everything is well planned and no parts are missing once on site.
- Weather conditions must be taken into account at the design stage. Zero output from the PV array from November through to February (1997–8) suggests that the PV array was covered with snow/ice; manual removal at such sites is practically impossible.

REMAINING ISSUES

Two aspects deserve particular attention:

- There is much room for optimizing sizing or control strategies in order to reduce expensive fuel consumption and run-time of generators.
- Rime accumulation can reduce PV array output at critical time, i.e. when energy is most needed; therefore a system with de-icing of the PV module would be beneficial.

PV electric propulsion

APPLICATION

A large market exists for electric drives for ships in a variety of sizes and uses, ranging from privately owned leisure boats and sailing ships up to professionally exploited ferries and excursion ships. Now that several natural areas are being closed for internal combustion engines, electric boating is becoming popular with a growing group of people. In many cases it has been shown that PV can supply sufficient energy to power these boats. This opens market opportunities for well-designed and optimally integrated PV generators on board of more than a million ships in Europe, resulting in many hundreds of MWp.

Several projects aim at the research and development of PV/battery systems. Different types of use and demand are being demonstrated currently on about a hundred systems in The Netherlands and Germany, with financial support of the EC-Thermie programme (se 60/94 NL/DE and dis 972/96 NL/BE/DE).

ADVANTAGES

The main advantages of PV electric propulsion in comparison with internal combustion engines are:

- No air pollution due to local emissions of greenhouse gasses;
- No noise pollution, thus preservation of natural areas;
- No water pollution through spilling of fuel;
- Access to areas otherwise closed to boats with internal combustion propulsion.

MARKET

- In The Netherlands 144 000 boats have the potential to be propelled by PV electric engines.

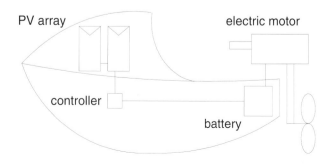

Solamaran with a 318 Wp PV system for electric propulsion

Lack of space available for PV modules sometimes causes problems.

Flexible flat modules have proven to be useful for PV integration in existing boat design.

- The technical potential is over 50 MWp.
- Commercial potential in The Netherlands is estimated to be 150 kWp/year.

The following usage groups contribute to the above-mentioned projections:

Usage group	Capacity (Wp)
Sailing boats (auxiliary propulsion)	150–250
Motor boats (main propulsion)	200–1 000
Large boats for professional use (e.g. ferries, excursion boats)	1 000–5 000

Cost estimation of a 200 Wp system (US $)

Component/ element	Present costs	Costs in 2005
PV modules	1 600	700
Batteries	400	400
Controller	200	200
Motor (500 W)	700	600
Cabling etc.	100	100
Installation	800	800
Gross system costs	3 800	2 800
Commercial margin 40%	1 520	1 120
Net system costs	**5 320**	**3 920**

INNOVATIVE ASPECTS

- Combination of electricity generation and electric propulsion;
- Integration of modules in existing boat design and production process;
- Energy conservation through boat and propeller design.

FINANCIAL ASPECTS

For newly-built ships, PV electric propulsion is economically feasible. Costs are comparable to propulsion by means of internal combustion engines. However, a number of items offer opportunities for cost reduction:

- Production process: integration of PV in hulls of new ships through lamination;
- Standardisation of components, e.g. small PV panels;
- Energy conservation in propulsion system (propeller adapted to low-speed rotation);
- Twinning of motor and PV suppliers.

LESSONS LEARNED

- The technology of the autonomous PV power supply is available and is satisfying most users.
- With respect to sizing, potential users tend to overestimate their energy demand.
- Flexible modules are appropriate, given the fact that one can walk over them.
- End-users should be thoroughly instructed on battery use and maintenance.
- Most leisure users actually navigate around 2 hours per day.

REMAINING ISSUES

The following aspects can still be improved upon:

- Corrosion of battery terminals.
- Improved battery management through regulators equipped with state-of-charge indicators.
- Limited space restricts PV modules: further integration of cells and other components into the boat design could overcome such limitations. A step further ahead is that boat design anticipates PV integration.

Photovoltaic parking meters

Portugal

APPLICATION

Parking meters are an essential part of urban management in the increasingly congested town centres. Parking meters encourage short-stay parking with high turnover, improve traffic conditions and promote trading activities.

In 1995 the municipality of Lisbon and EMEL, a municipal enterprise exploring parking places, started to install PV parking meters. The result of this policy is that today all parking meters in Lisbon are running on PV.

The load of a parking meter is mainly the coin machine and a printer. These are actually small loads and PV systems are quite adequate for this application.

TYPICAL SYSTEM CONFIGURATION

A typical parking meter represents a small load operating at 12 V. When in stand-by operation the load current is

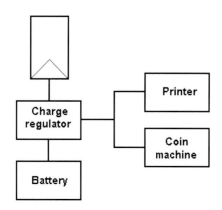

around 1 mA and a typical operation has a consumption of around 2 A for an average time of 5 s. The average number of daily operations is 100 with a maximum value of 300. This represents an average daily load of 3.6 Wh and a maximum of 10.3 Wh.

Taking into account that the worst month in Lisbon has an average of 2.75 hours of sun peak, a PV module of 5 Wp can supply the maximum load.

Typical batteries used are of the sealed lead-acid type with nominal capacities ranging from 12 Ah to 50 Ah.

ADVANTAGES

- A PV parking meter needs no grid connection, is easy to handle and can be placed anywhere.
- Combined with flexible software programming, parking meters can easily be adapted to operate under any conditions.
- These are highly reliable systems with low maintenance requirements.
- From the point of view of economics PV parking meters are a good choice in the long run when compared with grid-connected parking meters. They are without doubt a better choice when compared with parking meters running on dry batteries.

INNOVATIVE ASPECTS

Integration of all components in one makes installation and removal fast, easy and cheap.

Both the ticket printing and the coin mechanism have been designed to consume minimal energy.

Parking meters with integrated PV panels are less prone to theft.

Parking meter with a separate PV panel on a multi-functional pole.

MARKET

In Lisbon all parking meters, over 830, are photovoltaic. Also other cities are installing PV parking meters and today nearly 60% of all parking meters in Portugal are running on PV. This represents an installed power around 7 kWp. The number of PV parking meters in Portugal is expected to increase to 5 000 (around 40 kWp) in the coming years. This is due not only to the successful experience obtained in Lisbon, but also to the general trend to install parking meters in the city centres.

FINANCIAL ASPECTS

The average cost of a PV parking meter is in the order of 5 000 ECU, which is typically 10% higher than a conventional system.

LESSONS LEARNED

The installation of PV parking meters in Lisbon turned out to be a success mainly due to:

- Easy installation without the need to install underground cables;
  - Low maintenance;
  - High reliability.
- Care must be taken, when installing these systems, with the site choice and module orientation. In some cases basic rules for module installation were sacrificed to architectural and urban-planning considerations.
- Problems of vandalism were reported by some of the installers.

REMAINING ISSUES

- Battery capacity design, which in some cases was reported to be low. This can be related to a high number of operations or a high self-discharge rate of batteries.
- New functions like electric pay and telemetry are increasingly being incorporated in parking meters. Design of PV parking meters should anticipate this development and cope with the increased power demand.

Bus shelters

APPLICATION

The purpose of the PV system is to provide illumination at a bus stop for safety, for reading timetables and for the bus driver to know whether he should stop. An infrared detector automatically turns on the light when someone enters the shelter. The PV system is either installed in existing bus shelters or sold together with the shelter in the case of new installations.

TYPICAL SYSTEM CONFIGURATION

- A typical system consists of one 50 Wp module;

PL light and sensor, which is protected with a metal grid.

- 12V, 25 Ah lead-acid battery;
- 11 W fluorescent lamp
- Infrared detector reacts on movement and switches on the lamp, protected with metal grid.
- The daily energy consumption is 60–100 W.

ADVANTAGES

The main advantage is the low cost and simple installation because of the freedom from the utility grid.

MARKET

The first 400 systems are installed in the city of Gothenburg but the market might be bigger for the countryside where distances to the grid are longer and background illumination from streetlights is less.

Until now, systems have only been installed in the southern part of Sweden and no efforts have been made to open up the market in the north of the country.

INNOVATIVE ASPECTS

To reduce vandalism, all crucial components are mounted on a 3-metre high pole (see main photo). This also allows better orientation of the modules and avoids shading from surrounding objects in winter.

This figure shows how close the grid can be located. The grid is available both in the nearby houses and at the streetlight to the right of the bus-shelter. It is very expensive to dig trenches and then replace the pavement in a city. There is also a big risk of damaging other installations in the ground. Even just a short distance of a few meters can have a total cost of equivalent to US$5000.

FINANCIAL ASPECTS

- The cost of the PV system is US$ 1 300 and for a whole shelter US$ 8 000. Avoiding the grid connection costs (US$ 3 000–4 000) makes this application cost-effective from the beginning.
- Although normal running costs are minimal, replacement of stolen or destroyed components turns out to be costly. Solving this problem would be the main strategy for cost reduction.

LESSONS LEARNED

- The maintenance requirement has been mainly repair of lamps and infrared detectors, and installation of new modules due to theft.
- Vandalism and theft form a major problem, not only for PV-powered bus shelters but also for the grid-connected ones. In the first year, 80 modules have been stolen, as well as lamps and the infrared sensors.

Close-up of the bus shelter showing the lamp inside, the PV-module with battery and controller located just beneath on the pole. This is an attempt to reduce problems with theft and vandalism

REMAINING ISSUES

- Better integration of components should make them less prone to theft and vandalism.
- Special colours for the frame or etching of the glass cover could also contribute to theft reduction.
- When implementing the system in the northern parts of Sweden, the size of the panels and batteries will have to increase.

PV electric rail and flange lubricators

APPLICATION

PV-powered electric rail greasers prevent fatigue and noise on railway tracks.

TYPICAL SYSTEM CONFIGURATION

There are two possible system designs; the most popular of which is shown in the diagram below.

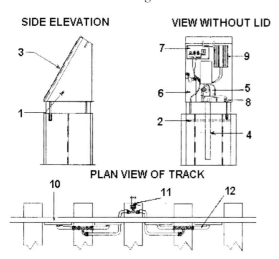

1   Cabinet moulded to support PV panel at a 60° angle to the horizontal. Contains pump, drive unit and electronics.

2   Ground tank holds a 25-kg grease keg.
3   Polycrystalline PV panel, complete with blocking diode.
4   Positive-displacement high-efficiency grease pump.
5   Motor/gearbox drive with 12 V DC permanent-magnet motor.
6   Dry-fit 12 V DC 15 Ah capacity battery.
7   Control unit with timers and switches for isolation and priming.
8   Grease outlet pipe from pump head to outside of cabinet terminating in hose tail connected to grease distribution system.
9   Voltage shunt regulator to avoid overcharge.
10 Four short grease distributor units.
11 Main hose feed from cabinet.
12 Short delivery hose with regulator valve.

ADVANTAGES

The main advantage of the solar-powered unit is that it can be easily re-sited. Another benefit is that installation can be carried out by the permanent track engineer, rather than having to involve external electrical contractors.

Two different sizes are available:

• 20 Wp model with 25 kg grease keg.
• Larger 50 Wp models for track-side mounting and storage of two 50 kg grease kegs.

Large model with two 50-kg grease kegs

MARKET

There is a potentially a very large world market for the PV powered rail greaser. This is already a very popular alternative to the battery-powered model in the UK, and the established European market is continually growing.

INNOVATIVE ASPECTS

- The PV panel is integrated into the product.
- Special control unit designed for this application.
- The low-voltage DC power is more appropriate than the mains power already available on the railway.

FINANCIAL ASPECTS

The PV-powered rail greaser has proved itself to be economically competitive with the battery- and mains-powered versions.

The product has been cost optimized. Reduction in PV panel price will further decrease costs.

Orientation of the lubricators is sometimes problematic...

Lubricator with the grease delivery hoses clearly visible.

LESSONS LEARNED

The PV-powered rail greaser is often the preferred option due to its light weight, compact size and the corresponding compliance with EU Health and Safety guidelines for manual lifting.

PROBLEMS TO BE SOLVED

In some cases, use of the product by untrained staff could result in misalignment of the PV panel, for example in a north-facing direction.

REMAINING ISSUES

Modifications in the product design would be required for applications outside Europe due to the difference in optimum angle of the PV panel.

PV electrification of summer cottages

Finland

APPLICATION

The system provides power to remote summer cottages, with typically low load levels. The cottages are mainly used in summertime, so the availability of solar energy coincides with consumption. Any heating requirement is usually satisfied with firewood. A typical load is two to three lights, colour TV, radio, and water pump.

TYPICAL SYSTEM CONFIGURATION

A typical system configuration is as follows:

- One module (50 Wp), occasionally two (100 Wp or more). The typical design insolation for summertime is around 5 kWh/m²/day.
- One 12 V 100–120 Ah flooded lead–acid battery for each module installed, pasted plates. At least 6 days of autonomy is recommended.
- ON/OFF series-type controller. Also more advanced devices with SOC calculation and DC/DC converters suitable for constant voltage charging as these emerge into the market.
- For systems with one module and one battery, the suitable load are two to three lights, water supply pump, and colour TV (used 2 h daily). The total daily consumption is then approximately 130 Wh. Larger systems with two batteries and three modules can power more lights and a low-consumption refrigerator.

ADVANTAGES

This PV application competes with grid extension. In most cases, the distance to the electrical grid is so long that PV is the most cost-effective option. Gasoline gensets can be considered as complementary devices, as people do not want to have them running all the time because of noise and pollution.

MARKET

There are now 16 000 to 20 000 summer-cottage PV systems in Finland, and around 150 000 summer cottages without electricity. It is difficult to estimate the actual potential of PV systems, as these are holiday homes and the

A summer cottage with a 2.2 kWp A-Si array.

User-friendly controller GENIO by NAPS.

Example of bad module siting

investment is therefore not directly related to the basic needs. However, it is obvious that there is still a large market.

INNOVATIVE ASPECTS

A main problem used to be overloading of the battery, thus shortening its lifetime due to sulphating and freezing. Reasons are underestimation of the load and load increase due to new appliances.

Hence, more advanced controllers have been developed that indicate battery state of charge and remaining operating hours at the present load. To prevent freezing of the battery, these regulators also adjust the control limits on the basis of the time of year and temperature. The MPPT maximizes energy yield from the PV modules, especially at low battery SOC, and enables battery-friendly constant-voltage charging.

FINANCIAL ASPECTS

The price of the basic PV system kit is approximately US$ 1 000. The cost of a grid extension is typically several times this amount.

The cost reduction of the initial investment mainly depends on the price development of the PV modules. The maintenance costs, with the battery replacement as the major contributor, are the only expenses that the user can influence. These can be minimized by:

• Proper sizing of the system.
• Saving energy, e.g. by using PL lights.
• Adjusting the energy consumption to the insolation.

LESSONS LEARNED

• Even though it is not a major problem, shading of the PV modules is the most common cause of trouble. Trees usually surround Finnish cottages and it is not easy to find non-shaded sites. The few problem cases can create suspicion about the performance of PV in general.
• The malfunction of the hardware (electronics etc.) is a rather minor cause of problems.
• Regarding the non-technical aspects, the expectations of the energy yield may sometimes be unrealistic. Therefore, it is important to disseminate realistic information on the typical system performance and applicable loads. In general, more user training and information is required (adapted to retailers as well).

REMAINING ISSUES

Some technical suggestions to solve the problems mentioned above are:

• A simple geometrical device to find an unshaded site for the PV module (for retailers/consumers)
• A larger variety of accessories for attachment of the modules to a supporting pole to raise the modules to an unshaded location.

PV rural electrification in the Jura

France

APPLICATION

This application, situated in the Jura Mountains, is part of the French PV rural electrification programme, managed by the French utility (EDF) and partly supported by the ADEME and the European Commission (Transeuropeo project). A stand-alone PV system is installed on roofs of remote houses far away from the grid. The PV system (1 to 2 kWp) supplies DC high-efficiency appliances for all permanent needs as well as some AC equipment for occasional demands.

The programme is mainly financed by funds for previous grid extension (FACE), which appears to be more expensive than PV.

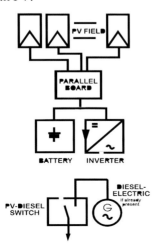

TYPICAL SYSTEM CONFIGURATION

A typical installation has 1.2 kWp, tubular plate batteries 800 Ah, no central inverter, DC lights, freezer, refrigerator, stereo, tools, plus AC washing machine, TV, etc.

ADVANTAGES

In the specific conditions of France (large grid supplying nearly everyone, cheap nuclear energy), PV is limited to remote and sunny areas such as mountains (Jura, Alps) or the south (Provence, Roussillon). PV is cheaper than grid connection and better for the environment (no cables and masts).

MARKET

- In France, the market is limited to 3 000 to 5 000 houses.
- In French overseas territories, the market is over 100 000 houses.
- In the coming 20 years, the French market could be extended to new dwellings in sunny areas, even if close to the grid. PV will enable houses to be built anywhere.
- This market depends of the availability of high-efficiency appliances. Such appliances will first be developed for the developing countries market.

Remote villa with PV system integrated in roof.

Isolated farm with small PV system.

Cost estimation of 1.2 kWp system (US $)

Component/ element	Present costs	Costs in 2005
PV modules	6 000	4 000
Batteries	4 800	4 200
Controller, inverters	2 800	1 800
Support structure	1 000	800
Specific appliances	3 000	1 500
Monitoring	1 400	200
Installation	8 000	5 000
Spare parts	1 000	700
Gross system costs	28 000	18 200

Costs are related to the installation of about 200 plants (240 kWp).

Inauguration of a PV system.

INNOVATIVE ASPECTS

The programme integrates lessons learned in overseas territories:

- Specifications: the *Charte de qualité* specifies all the aspects of the PV house: sizing, installation, integration, and management.
- High efficiency: the system minimises consumption and, in consequence, the size and the price of the PV generator.
- Maintenance (including replacement of batteries) is responsibility of the utility (EDF).

FINANCIAL ASPECTS

A stand-alone PV system is less expensive than grid extension if the distance from the existing grid exceeds 500–1 000 m for an average installation of 1.2 kWp.

Cost reduction will be the result of mass production for large markets such as electrification in developing countries. Of all components, the highest price reduction is expected from the controller and specific DC appliances, which are nearly non-existent to day.

LESSONS LEARNED

- AC or DC: no unique answer can be given today. While AC is a good transition, DC is better for small or very remote installation, as well as developing countries.
- Hybrid system: most installations are hybrid with small back-up diesel gensets.
- Maintenance: Batteries are the weakest element.

REMAINING ISSUES

- Barriers and user acceptance:
  - barriers from local authorities;
  - financial scheme for funding PV;
  - acceptance of energy restriction;
  - acceptance of DC instead of AC;
- Technical issues:
  - norms, standards, quality control;
  - high-efficiency DC appliances;
  - improved battery storage;
  - supply of professional demands.

PV–hybrid system with AC-coupling

Germany

APPLICATION

The alpine refuge Starkenburger Hütte is located in the Stubaier Alps at an altitude of 2 229 m. This alpine refuge is operating from beginning of June to late September. During the season the Starkenburger Hütte has approximately 1 600 sleeping guests and 6 000 day-guests. Kitchen equipment, lighting and washing facilities have to be supplied with electrical energy.

The Starkenburger Hütte is a pilot and demonstration project for the recently developed AC coupling technology: a hybrid system based on AC coupling of all devices including the battery.

TYPICAL SYSTEM CONFIGURATION

The system contains 5 kWp PV, a 14 kW gas combined heat and power generator and three AC batteries (BacTERIE) with a capacity of 10 kWh/2.2kW each.

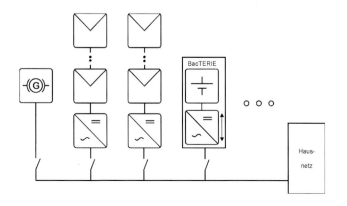

ADVANTAGES

- Standardization to 230 V AC / 50 Hz enables mass production and thus improvement in reliability and cost reduction.
- Applicability of standard 230 V AC / 50 Hz appliances.
- AC is suited to being implemented by traditional professionals.
- AC-coupled systems are easy to incorporate in future grid extensions. In that case, AC-coupled PV–hybrid systems contribute to stabilization of weak grids.

MARKET

- About 1 400 mountain refuges in the alpine region which are similar to Starkenburger Hütte.
- A huge potential can be seen worldwide: 2 billion people live without electricity, most of them in areas with a low population density where hybrid systems are the most economic alternative.
- In regions with higher population density extended grids are often weak; AC–PV Hybrid systems contribute to their stabilization.

INNOVATIVE ASPECTS

- The first PV–hybrid system with AC-coupled components.
- The so-called BacTERIE has been developed. This energy storage unit contains, inside a single cabinet, the batteries, a temperature equalising unit, an integrated battery management system and a bi-directional inverter; the interface is purely AC.

BacTERIE: battery unit with bi-directional inverter and battery management system

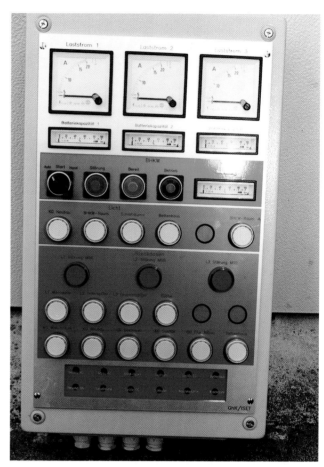

Control unit indicating system status and consumers in operation

There is an enormous potential for cost reduction as many components are prototypes developed especially for this project.

LESSONS LEARNED

- The system has been able to supply the refuge with electrical energy and heat in a secure and reliable way from its installation in July 1997 until the refuge closed in October.
- All AC components could be assembled and installed at the refuge site easily and in a short period of time.
- The operator didn't have any problems with the handling of the system. The control table giving information about the state of the system and the consumers is a big help for a rational energy use.

REMAINING ISSUES

- The AC technology is still in its childhood: extensive monitoring is recommended.
- Until the components are mass-produced the concept is too expensive without public funding.
- A synchronous three-phase system has to be developed.
- A standard AC energy bus is the obvious choice, although a lack in communication of components still exists.
- Interfaces and communication have to be standardized.

Outside view of the Starkenburger Hütte

- The refuge staff can check the statuses of both the system and the consumers.
- The colour design of the units should lead towards a standardization in using symbols and colours for marking AC, DC, and control elements.

FINANCIAL ASPECTS

It is not possible to give realistic figures concerning the costs of this pilot project, which was funded by the German Ministry of Education, Science, Research and Technology (BMBF) under contract number 0329549A.

Utility PV: 'Wireless Service'

Italy

APPLICATION

The PV system supplies remote users where the grid extension is not permitted by environmental law restrictions (parks, natural areas, etc.), or where the cost comparison (low-voltage grid-connection vs. PV system installation) benefits the utility's rural electrification programme and reduces the connection costs for the user.

The operation and maintenance are the responsibility of the utility that owns the system. The user as part of the energy supply contract provides the plot free of charge. The kWh cost charged to the user is the same as for a grid connection.

TYPICAL SYSTEM CONFIGURATION

The PV plant is equipped with an inverter for the AC electric distribution to the loads, thus allowing the use of commercial appliances.

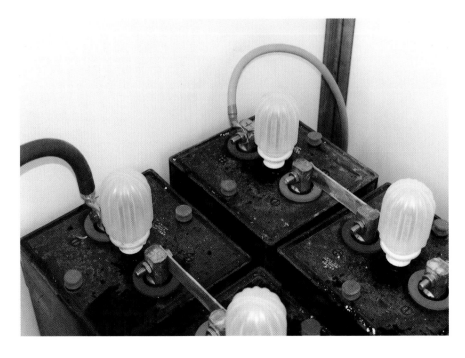

Special plugs reduce refilling needs.

PV system with proper fencing and system house.

ADVANTAGES

PV systems avoid the installation of the usual diesel gensets, save on fuel cost and fuel refilling and reduce operation noise. If already on site, the genset can be used as a back-up in case of PV failure or energy shortage.

MARKET

Despite the initial high capital costs, the PV supply of remote houses has high market potential because of the low O&M costs compared to diesel gensets. An investigation in Italy showed that expectations (including private and utilities rural programmes) stand at 5 000 systems with a 0.5% annual growth (theoretically 500–750 kWp/year).

Basic DC systems with optional small-size inverter for essential AC appliances can be designed using the same modularity criteria as for applications in developing countries.

INNOVATIVE ASPECTS

- Modularity: a range of system sizes (1.5–3 - 6 kWp) has been installed using the 750 Wp basic standard unit.
- Maintenance: batteries are equipped with special plugs to reduce the refilling with distilled water.
- Safety: the design adopted both the utility and international standards

FINANCIAL ASPECTS

Considering the average Italian kWh price and using current component prices, the PV plant is cheaper than grid extension if the distance exceeds about 700 m per kW required. This parameter has been presented in a matrix (distance to grid versus power requested) for quick evaluation by the utility.

Although the long-term total cost reduction is mainly dependent on the module price, a different layout (integration without fence) and BOS optimization offer most prospects for short-term cost reduction.

Control unit.

Cost estimation of 3 kWp system (US $)

Component/ element	Present costs	Costs in 2 005
PV modules	12 000	10 500
Batteries	9 000	7 500
Controller, inverters	6 300	4 500
Support structure	2 550	2 400
DC boards, wiring	4 200	3 000
Installation, civil work	5 850	1 500
Checks & start-up	600	600
Gross system costs	40 500	30 000

Costs are related to the installation of about 150 plants (450 kWp).

LESSONS LEARNED

- Reliability: in the first-generation inverters, the high efficiency target wasn't well balanced with the necessary reliability. The second-generation inverter designed by ENEL is built using essential criteria (inox shelters for outdoor installation, high-frequency commutation for noise reduction, and no electromechanical components). It has been successfully tested for two years under extreme environmental conditions.
- Maintenance: the device used for the battery storage reduces the refilling period to only once a year.

REMAINING ISSUES

- User acceptance:
  - financial scheme for funding PV;
  - acceptance of DC instead of AC;
  - can standardisation in sizes and layout really satisfy the market potential?
- Technical issues:
  - Better batteries;
  - AC or DC + small inverter?

PV–DG hybrid systems for remote dwellings

APPLICATION

This application is part of a programme to reduce the cost of the rural electrification, managed by the Norwegian Water Resources and Energy Administration. A stand-alone PV–DG hybrid system is installed to supply a remote dwelling house with electricity.

TYPICAL SYSTEM CONFIGURATION

Cottages have a system of typically 100–500 Wp, while tourist lodges range from 500–1 000 kWp. Full-time inhabited family houses need approximately 2 000 kWp.

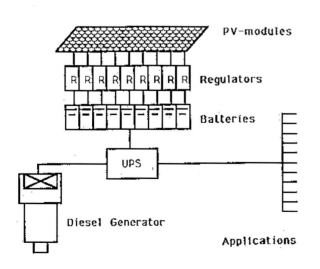

ADVANTAGES

The PV modules contribute to reduce the fuel consumption and the operating time of the DG in the summer season, thus contributing to noise reduction. It is also expected that the PV-modules charge the batteries more favourably and thus contribute to extend their lifetime.

MARKET

In Norway the number of family houses where such a local stand-alone power system can be an economic solution is estimated at 1 000–3 000, depending on the cost of the local system. One has to add to that an unknown number of remote cottages, tourist lodges and other isolated sites require high-standard energy services.

INNOVATIVE ASPECTS

- A PV generator in conjunction with a diesel generator at a relatively small scale.
- Energy management, battery use.
- Energy conservation.

FINANCIAL ASPECTS

A stand-alone power system becomes less expensive than the grid extension when the distance from the existing grid exceeds 2–3 km. A hybrid solution using a PV generator is economically feasible if the annual consumption is less than 3 000 kWh.

Detail of the PV array.

Cost estimation of a 2 kWp system (US $)

Component/ element	Present costs	Costs in 2 005
PV modules	15 000	10 000
Batteries	12 000	10 000
Controller, inverters	5 000	5 000
UPS	8 000	7 000
Diesel generator	10 000	-
Sterling generator	-	2 000
Generator house	15 000	15 000
Gross system costs	65 000	49 000

Possibilities for cost reduction are:

- Standardization of components and system design (one solution for many houses makes mass production possible).
- Energy conservation in appliances.

LESSONS LEARNED

- For a family that expects all ordinary facilities to be available in the house, an AC solution is necessary.

- Battery management must be an integrated part of the system and not just rely on the end-user.
- Utilities lack experiences in SAPV projects and need proper attention in order to cooperate.
- End-users need ample education to adopt their habits to limited energy supply.

REMAINING ISSUES

- Insufficient availability of DC household appliances.
- The DG is sometimes competing with the PV system. In these cases, the contribution from the PV array is less than expected.
- Remote monitoring of the system in order to detect operational disturbance at an early stage.
- Adaptable operational parameters due to changing conditions throughout the year.
- A cheaper Sterling engine could replace the diesel genset.

PV rural electrification with concerted service

Spain

APPLICATION

La Garrotxa (Catalonia, Spain) is a district with 735 km² and 46 060 inhabitants (1991). Electrification is very important for the development of such rural regions. SAPV provides new opportunities for a decentralized infrastructure supplying electrical energy to rural population, where low individual energy demands render extensive grid lines not feasible while operation of diesel gensets is expensive (inefficient at partial loads, costly fuel transport, high maintenance).

At 69 rural sites, 52.2 kWp have been installed. Specific requirements were drawn up for each site based on individual load estimates.

TYPICAL SYSTEM CONFIGURATION

The system layout is standardized and completely modular, including the power-conditioning unit:

- Typically 800 W PV modules, 19.2 kWh battery capacity.
- DC/AC inverter expandable in 1 kW steps up to 4 kW.

ADVANTAGES

- Advanced PV systems provide an equivalent service to grid extension and supply 230 V_{AC}/50 Hz (sine wave).
- A single operator ensures maintenance of the equipment.
- 160 km grid lines are avoided in vulnerable forested landscape.

MARKET

- High grid-extension costs (20 000 US$/km in flat and open terrain, more in forested or mountainous areas) make PV systems a clear lower-cost alternative.
- Two billion people worldwide need electrification in rural regions.

INNOVATIVE ASPECTS

- Operation, monitoring and maintenance of the PV plants are the responsibility of one single energy operator: a users' association. This association (SEBA, see Chapter 6), especially conceived for this purpose, guarantees quality of service with a 15 year contract.
- User fees are as in grid connections.

Roof-mounted PV system at a remote farm in the Garroxta region.

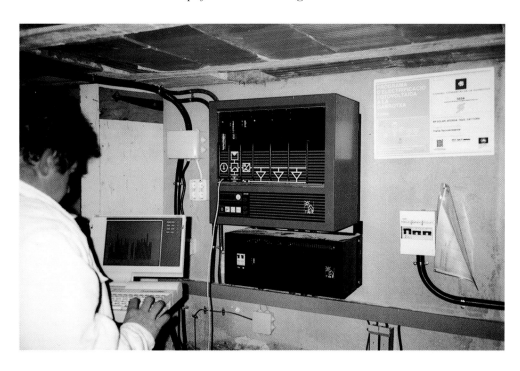

The data logger is integrated in the control unit.

Cost estimation of 1 kWp system (x 1 000 US $)

Concept	Present costs	Envisaged costs in 2005
PV array	4.8	4.0
Batteries (20 Wh/Wp)	2.3	2.0
Inverter (2.5kW/kWp)	1.9	1.3
MPP Tracker	0.5	0.3
Load management indicator	0.8	0.2
Structure	0.5	0.5
PV installation	2.5	2.5
SUBTOTAL PV-SYSTEM	13.3	10.8
Indoor wiring	2.8	2.8
High-efficiency loads	2.5	2.5
Back-up LPG generator	2.5	2.5
SUBTOTAL IMPLEMENTATION	7.8	7.8
Service, follow up	1.4	0.8
Monitoring design	0.3	0.3
Data gathering/analysis	1.0	0.2
SUBTOTAL AFTER SALES	2.7	1.3
Project administration	0.4	0.3
Legal counselling	0.3	0.1
Maintenance structure	0.5	0.4
Commissioning. reporting	1.0	0.7
Programme management	1.0	0.7
SUBTOTAL ORGANISATION	3.2	2.2
TOTAL	**27.0**	**22.1**

- Users trained to contribute to maintenance, to take care of load management and to use high-efficiency appliances.
- Standardized equipment components.
- Modular concept, also including power conditioning.

FINANCIAL ASPECTS

- Financing of the system: individual users with the support from local rural electrification programs and the THERMIE program of the EU (SE/084/92 ES).

- Maintenance costs reduced by three level organisation:
  - User: inspection of operating parameters and electrolyte level.
  - Local technician: preventive maintenance, repairs and data collection.
  - Engineering consultants: data analysis; feedback on available energy; maintenance management.

LESSONS LEARNED

- Standardization and modularity reduces installation time and design costs.
- High satisfaction among the users.
- User participation and training reduces maintenance cost and contributes to a general awareness of the rational use of energy.
- Demonstration to local bodies subsidizing grid-extension programmes helps in convincing them that lower-cost alternatives are available and provide equivalent service.

REMAINING ISSUES

- Access to public funds for rural electrification has to improve.
- Comprehensive schemes to provide PV energy have to be further developed.
- Integration of SAPV with other energy services (solar hot water, LPG, etc.).
- Power conditioning: integrated high-efficiency equipment with emphasis on replacement of parts and servicing.
- User-friendly interfaces for energy management.
- High-efficiency appliances.

PV–DG hybrid systems for remote islands

Japan

APPLICATION

In most remote islands, electric power has been provided traditionally by small- or medium-sized diesel generators (DG). However, DG have a number of problems, such as high power-generation cost and unstable fuel transport. In order to reduce these problems, the New Energy & Industrial Technology Development Organisation (NEDO) developed the practical application of PV–DG hybrid systems for remote islands, as a part of the New Sunshine Program of the Ministry of International Trade and Industry (MITI).

In this system, the diesel generator only provides power when the PV arrays generate insufficient power because of poor insolation. When the generator runs, it also recharges the storage batteries, through an inverter, thus improving its load characteristics and hence reducing the fuel consumption.

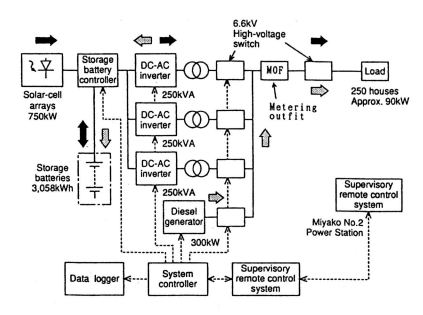

System configuration.

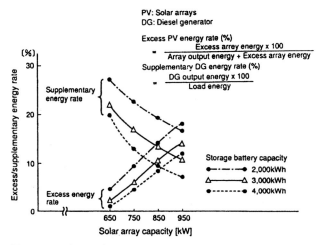

PV: Solar arrays
DG: Diesel generator

Excess PV energy rate (%)

$$= \frac{\text{Excess array energy} \times 100}{\text{Array output energy} + \text{Excess array energy}}$$

Supplementary DG energy rate (%)

$$= \frac{\text{DG output energy} \times 100}{\text{Load energy}}$$

Excess and supplementary energy rates.

Aerial view of the 750 kW PV array.

TYPICAL SYSTEM CONFIGURATION

The system has:

- 750 kWp of PV;
- 3 058 kWh of batteries; and
- 300 kW diesel generator.

ADVANTAGES

The PV–DG hybrid system has many advantages:

- More efficient utilization of PV power than the stand-alone PV system;
- Higher reliability of power supply;
- Lower demands for PV cell and battery capacity, as illustrated in the table below;

System	Hybrid	SAPV
PV cells (kWp)	750	1 200
Battery (kWh)	3 058	13
DG (kW)	300	–
System utilization (%)	10.5%	7.7%

Average load: 90 kW, annual irradiation: 1 598 kWh/m²

Battery recharging by DG reduces the fuel consumption by about 20% in comparison to a DG system.

MARKET

- This system is the first large-scale hybrid system in Japan.
- There are 61 remote islands in Japan provided with power supply systems with a total energy production of 1 722 GWh (1988). The PV–DG hybrid system is expected to be introduced in 22 small islands with a system capacity of 500 kW or less. Total energy production in the latter category is 12 GWh. The estimated market size is about 12 MW.
- In the unelectrified areas of south-east Asia, the PV–DG hybrid system is expected to be used also for village power supply.

FINANCIAL ASPECTS

System costs: about US $ 16 million, variable depending upon conditions.

Options for cost reduction:

- Optimization of system capacity.
- Optimization of array inclination parameters (power generation, construction cost, installation area).
- Control on inverter side (saving devices for DG, such as automatic frequency control (AFC), synchronized parallel connection, and load share control).

Cost estimation for 750 kWp PV + 300 kW DG (Million US $)	
Component/ element	Present costs
PV modules	5.38
Supporting structure	1.34
Batteries	1.36
Power conditioner	1.07
System controller	0.43
Diesel generator	1,00
Construction work	5.42
Gross system costs	**16.00**

LESSONS LEARNED

- Experience has been gained in AC parallel operation technology for PV and DG.
- Insight into system capacity and generation characteristics (PV surplus, DG supplement, power factor).
- The mean efficiency can be improved by about 3% through adjustment of the number of active inverter units.

REMAINING ISSUES

- Improvement of PV cell insolation under the high salinity environment typical of remote islands.
- Size reduction, efficiency improvement and cost reduction of the system equipment.
- Extension of battery service life.

Remote monitoring of PV–diesel hybrid systems

Korea

APPLICATION

The Korean Government, through the Rural Electrification Project, intents to provide electric power under utility standards to remote islands. Traditionally, this is done through diesel generators, but now efforts are concentrated on finding alternatives with higher reliability and lower maintenance requirements. For small islands with a population below 50 households, a solution is found in autonomous PV–diesel hybrid systems with a multi-channel remote monitoring system. This technology has now been applied on three islands. The KIER (Korean Institute of Energy Research) is responsible for the operation and maintenance of the systems.

TYPICAL SYSTEM CONFIGURATION

A typical example is the Hawhado Island:

- Initially, 24 kWp was installed for 48 households. Although a considerable part of the population had left as a result of unemployment in the fisheries sector, after a few years the system had to be extended to 60 kWp to cover the growing energy demand of the remaining 34 households.
- The battery storage consists of 1 148 kWh lead–acid cells.

- A power controller of 80 kW and two inverters of 25 kW complete the PV system.
- Two generators of 75 kW and a rectifier of 100 kVA provide the back-up system.
- The monitoring systems consist of an on-site computer that collects analytically all important parameters of the system and displays these in real time. The data are transferred by telephone line to the KIER institute, where the central computer stores and displays data of several hybrids from all over the country. Warnings are issued in cases of abnormal and emergency conditions.

ADVANTAGES

- The remote monitoring system guarantees reliable, convenient and cheap maintenance with few qualified staff.
- The PV system reduces fuel consumption, resulting in less pollution and diminishing risks of oil effluence accidents during transport.
- Absence of noise.

MARKET

- Current government plans are to equip 102 non-electrified islands, having a total of 2 000 households, with PV–diesel systems. In the coming years, the total PV installed capacity is expected to surpass 5 MWp.

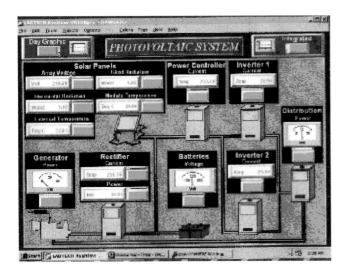

System layout as presented in the remote monitoring program.

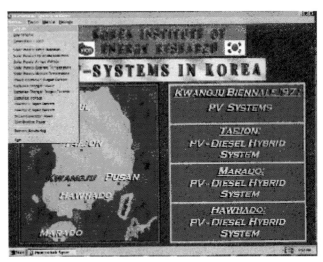

Opening screen of the remote monitoring program.

- This system is also expected to find its way to remote villages without any prospect of electrification through other means.
- Worldwide, thousands of comparable islands could be electrified in this way.

INNOVATIVE ASPECTS

The interactive user-friendly monitoring software was specially developed for this application.

FINANCIAL ASPECTS

For islands with less than 50 households, the initial investment of a hybrid system is comparable to that for a diesel genset. Its operating cost is about 70% lower. An additional 10% cost reduction can be obtained through the remote monitoring system.

The running costs of the system are heavily subsidized: although the annual expenses rise to US$ 35 000, only US$ 3 000 of fees are collected. Labour expenses account for the bulk of the O&M costs, while the diesel consumption is limited to a mere US$ 1 000.

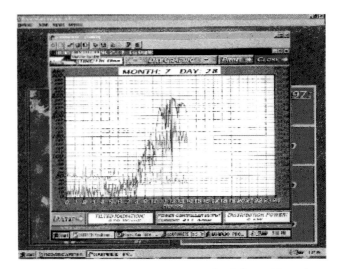

All important system parameters can be followed hour by hour...

LESSONS LEARNED

In order to increase battery life by avoiding deep depth of discharge (DOD), it is advisable to dimension the battery bank so as to cover three sunless days without diesel generation.

The remote monitoring system halved the O&M costs compared to a 'normal' PV system, while the overall running costs are no more than 20% of the diesel alternative.

REMAINING ISSUES

- Stability increase through improvement of surge current absorption from external loads and lightning.
- Increased battery lifetime.
- Improved resistance of the monitoring system to telephone line interruptions.

Cost estimation of 60 kWp Hawhado system (1 000 US $)

Component/ element	Present costs	Costs in 2005
PV modules	300	200
Batteries	100	90
Controller, inverters	100	90
2 Diesel gensets	30	30
Operation house	154	154
Transmission line	60	60
Engineering costs	130	100
Software development	34	2
Gross system costs	908	726

References

1 M.S. Imamura, P. Helm and W. Palz, Commission of the European Communities, *Photovoltaic System Technology, a European Handbook*, 1992

2 G. Loois (ed.), *Examples of Stand-Alone Photovoltaic Systems*, IEA Task III 1995

3 *Photovoltaics in 2010*, European Commission, DG Energy, Luxembourg, 1996

4 *PV News*, February 1998, pp4–5.

5 *Photovoltaics in 2010*, European Commission, DG Energy, Luxembourg, 1996

6 B. Chabot, M. van Brummelen, D. de Jager., *Large Scale Market Deployment of Photovoltaic Technology in Europe*, Ademe, ECOFYS, ETSU, ENEL, IEH, Sofia Antipolis, France, 1997

7 *Photovoltaics: On the verge of Commercialisation*, UPVG, Washington DC, June 1994

8 G. Loois, J. Verschelling, *Strategy for Market Introduction of Stand-Alone PV Systems in The Netherlands*, ECOFYS, Utrecht, March 1997

9 W. Palz, Power for the world: A global action plan, in *Yearbook of Renewable Energy 1994*, James & James, London, 1994

10 A. Cabraal, *Best Practices for Photovoltaic Household Electrification Programs*, ASTEA, World Bank, Washington, 1995

11 G. Loois, Bennalou, *Feasibility Study for Solar Energy Systems for Rural Households in the Rif in Morocco*, ECOFYS, The Netherlands, June 1994

12 *A Solar Wireless Electric Service, Electrical energy supply by means of photovoltaic power systems to remote off-grid users*, ENEL, Italy, 1995

13 X. Vallvé, J. Serrasolses, Operation experience of the 51 kWp stand-alone PV electrification programme in La Garrotxa (Catalonia, Spain), 13 EPVSEC, Nice, 1995

14 *Examples of Stand-Alone Photovoltaic Systems*, IEA Task III, 1995.

15 K. Burges, G. Loois, Van der Weiden, *Market Potential and Introduction of Hybrid Stand-Alone Power supply in Europe*, ECOFYS for EC DG XVII, April 1998

16 The study is produced by ECOFYS in close cooperation with utilities (EDON, NUON, REMU, EnergieNed), PV industry (associated in Holland Solar), ECN and Novem. As a fol-low-up, this ambitious strategic plan is supported by all above-mentioned sectors (utilities, PV industry, research, and authorities).

17 J. Gregory, S. Silveira, A. Derrick, P. Cowley, C. Allinson, O. Paish, *Financing Renewable Energy Projects, A Guide for Development Workers*, IT-publications, Nottingham, 1997

18 G.F. Bakema, A.P. Lentz, The utility supplies PV power, *Dutch Solar Energy Conference*, Veldhoven, 1995

19 ADEME, EDF, *Specifications for the Use of Renewable Energies in Rural Decentralised Electrification*, 1997-10

20 P. Varadi, Global approval programme for photovoltaics (PV GAP), *14th European Photovoltaic Energy Conference*, Barcelona, 1997

21 *Universal Standards for Solar Home Systems*, EC DG XVII contract SUP-9956-96 and EC DG XII contract JOR3-CT98-0275.

22 X. Vallvé, J. Serrasolses, PV stand-alone competing successfully with grid extension in rural electrification: a success story in Sotherm Europe. *Proceedings of 14th European Photovoltaic Solar Energy Conference*, 1997, pp. 23–26

23 X. Vallvé, J. Merten, E. Figuerol, Innovative load management for multi-user PV stand-alone systems. *Proceedings of 15th European Photovoltaic Solar Energy Conference*, Vienna, July 1998.

24 M. Ross, J. Royer eds., *Photovoltaics in Cold Climates*. James & James (Science Publishers) Ltd, 1998

25 D. Spiers, J. Royer, *Guidelines for the Use of Batteries in Photovoltaic Systems*. Neste Advanced Power Systems, Helsinki Finland and Energy Diversification Research Laboratory, CANMET, Department of Natural Resources Canada, Varennes, Canada, 1998

26 K. Burges, T.C.J. van der Weiden, K.J. Hoekstra, *Terschelling Hybrid System Research – Final Report*. ECOFYS, Utrecht, 1995

27 K. Burges, G.M. Koerts, G. Loois, *Data-Acquisition and Performance Evaluation for Autonomous PV Applications*. ECOFYS, Utrecht (NL), 1997

28 B. Andersson, S. Ulvonas, *How to Operate Solar Cell/Battery Systems*. Catella Generics Centre of Battery Technology, Stockholm, Sweden. 1997